Agriculture Issues and Policies

www.novapublishers.com

Agriculture Issues and Policies

Computational Intelligence for Sustainable Development
Brojo Kishore Mishra, PhD (Editor)
2022. ISBN: 979-8-88697-198-9 (Hardcover)
2022. ISBN: 979-8-88697-346-4 (eBook)

Pest Management: Methods, Applications and Challenges
Tarique Hassan Askary, PhD (Editor)
2022. ISBN: 979-8-88697-268-9 (Hardcover)
2022. ISBN: 979-8-88697-393-8 (eBook)

Jute: Cultivation, Properties and Uses
Matthieu Issa (Editor)
2022. ISBN: 979-8-88697-490-4 (Softcover)
2022. ISBN: 979-8-88697-505-5 (eBook)

Diseases of Fruit and Plantation Crops and Their Sustainable Management
Mujeebur Rahman Khan, PhD
Ziaul Haque, PhD (Editors)
2022. ISBN: 978-1-68507-978-9 (Hardcover)
2022. ISBN: 979-8-88697-297-9 (eBook)

Strategies to Achieve Sustainable Development Goals (SDGs):
A Road Map for Global Development
Rajani Srivastava, MSc, PhD (Editor)
2022. ISBN: 978-1-68507-836-2 (Hardcover)
2022. ISBN: 979-8-88697-027-2 (eBook)

Pistachios: Cultivation, Production and Consumption
Shaziya Haseeb Siddiqui, PhD
Shoaib Alam Siddiqui, PhD (Editors)
2022. ISBN: 978-1-68507-949-9 (Hardcover)
2022. ISBN: 979-8-88697-063-0 (eBook)

More information about this series can be found at https://novapublishers.com/product-category/series/agriculture-issues-and-policies/

Suryaprabha Deenan
Satheeshkumar Janakiraman
Seenivasan Nagachandrabose

Principles of Digital Image Processing for Agricultural Applications

www.novapublishers.com

Copyright © 2023 by Nova Science Publishers, Inc.
DOI: https://doi.org/10.52305/LZGO5867

All rights reserved. No part of this book may be reproduced, stored in a retrieval system or transmitted in any form or by any means: electronic, electrostatic, magnetic, tape, mechanical photocopying, recording or otherwise without the written permission of the Publisher.

We have partnered with Copyright Clearance Center to make it easy for you to obtain permissions to reuse content from this publication. Simply navigate to this publication's page on Nova's website and locate the "Get Permission" button below the title description. This button is linked directly to the title's permission page on copyright.com. Alternatively, you can visit copyright.com and search by title, ISBN, or ISSN.

For further questions about using the service on copyright.com, please contact:
Copyright Clearance Center
Phone: +1-(978) 750-8400   Fax: +1-(978) 750-4470   E-mail: info@copyright.com.

NOTICE TO THE READER

The Publisher has taken reasonable care in the preparation of this book, but makes no expressed or implied warranty of any kind and assumes no responsibility for any errors or omissions. No liability is assumed for incidental or consequential damages in connection with or arising out of information contained in this book. The Publisher shall not be liable for any special, consequential, or exemplary damages resulting, in whole or in part, from the readers' use of, or reliance upon, this material. Any parts of this book based on government reports are so indicated and copyright is claimed for those parts to the extent applicable to compilations of such works.

Independent verification should be sought for any data, advice or recommendations contained in this book. In addition, no responsibility is assumed by the Publisher for any injury and/or damage to persons or property arising from any methods, products, instructions, ideas or otherwise contained in this publication.

This publication is designed to provide accurate and authoritative information with regard to the subject matter covered herein. It is sold with the clear understanding that the Publisher is not engaged in rendering legal or any other professional services. If legal or any other expert assistance is required, the services of a competent person should be sought. FROM A DECLARATION OF PARTICIPANTS JOINTLY ADOPTED BY A COMMITTEE OF THE AMERICAN BAR ASSOCIATION AND A COMMITTEE OF PUBLISHERS.

Additional color graphics may be available in the e-book version of this book.

Library of Congress Cataloging-in-Publication Data

ISBN: 979-8-88697-428-7

Published by Nova Science Publishers, Inc. † New York

Contents

Preface		vii
Chapter 1	Introduction to Digital Image Processing	1
Chapter 2	Various Applications of Digital Image Processing	11
Chapter 3	Role of Image Processing in the Agriculture Sector	21
Chaptaer 4	An Overview of Fundamental Digital Image Processing Procedures	41
Chapter 5	Image Acquisition Methods Appropriate for Various Agricultural Situations	51
Chapter 6	Image Enhancement Tools for Images Acquired in Different Agro Conditions	57
Chapter 7	Image Segmentation Methods Concerned with Plant and Agriculture Images	77
Chapter 8	Feature Extraction Techniques Applicable for Agricultural Images	123
Chapter 9	Image Classification or Pattern Recognition to Solve Agricultural Problems	131
Chapter 10	Quality Assessment Metrics of Reference and Non-Reference Agriculturally Important Images	135
Chapter 11	Case Study on Successful Digital Image Processing Application on Banana	151
References		159
Index		167
About the Authors		171

Preface

Agriculture is a significant contributor to society as it forms the base for the survival of any living organism. The traditional agriculture system has evolved through various revolutions like the green and white revolutions. Recent advances in science and technology paved a new evolutionary era where machines and computer technology play a vital role. Computer science is now contributing to have more significant impact in the agriculture sector and is facilitating the farmers to ease their job. Computer technologies like machine learning, computer vision, image processing and machine vision have more scope and applicability in the agriculture sector. Digital image processing is a branch of computer science that now occupies a concrete position in the health sector for disease diagnosis and industrial sectors for quality control of products. However, its role in the agriculture sector is still infant stage. Digital image processing has several applications in agriculture sector like Global Positioning System (GPS) based decision support systems; identification of plants; soil quality analysis based on soil color; seed quality checking; pest and disease diagnosis and plant health monitoring; to ensure crop maturity; robot harvesting; grading; quality control of agro-products; and to solve so many problems in agriculture.

Due to the higher success ratio of digital image processing in other sectors, the agriculture sector has initiated to work in combination with the innovative information technologies like computer vision, machine vision and robotics, which could mimic the decisions of subject matter specialists. These techniques solve the input image defects acquired through imaging tools like digital cameras, sensors, thermal imaging and hyper spectral imaging. Image processing methods like image enhancement, image segmentation and feature extraction are applied over the input image to output required information that supports the farmers in decision-making. Digital image processing is more informative and supportive for farmers, agro-based industries and marketers. It can provide timely support for decision-making at an affordable cost.

Digital image processing has emerged as a separate course curriculum in biological sciences and is taught in many Agricultural Universities worldwide. Digital image processing is now offered as an independent course in many agricultural universities in India, USA, and other developing and developed countries. Several books on digital image processing target engineers as the primary audience. However, fundamental and easy-to-understand books on digital image processing for agricultural workers are scanty or nil. The driving force for this book is to fulfill the requirements of agricultural students and researchers who study or offer digital image processing in State Agricultural Universities in India and USA. Many students, scholars and scientists in the State Agricultural Universities need introductory textbooks on digital image processing to solve agricultural field problems. This basic, easily understandable digital image-processing book also appeals to agricultural extension workers and innovative farmers.

This study book discusses various applications of digital image processing in the agriculture sector and basic steps of digital image processing like image enhancement, image segmentation, feature extraction, and pattern recognition. Various algorithms available for image enhancement, segmentation, feature extraction, and pattern recognition are dealt with in this book with suitable problem-solved agricultural images. The quality analysis indices used for various kinds of image output are detailed. The book also has a chapter in which brief success stories on banana pest and disease diagnosis, fruit maturity and grading using digital image processing protocol. We hope this book will fulfill the requirements of agricultural graduate students, scholars, scientists, extension workers and lead farmers.

Chapter 1

Introduction to Digital Image Processing

Digital image processing is an inevitable application for image analysis with recent innovations in science and technologies. Invisible or uncertain objects can be made visible by the use of image processing. In general, image processing is the processing of digital images using a digital computer. The digital image represents a two-dimensional image as a finite set of digital values, called picture elements or pixels. These digital (discrete) values are operated to test the 2D space on a regular grid and to quantize each model (round to nearest integer). The samples that are 'D' apart in a discrete image are expressed as, $f[i,j] = Quantize\{f(i\,D, j\,D)\}$ where 'D' is distance, 'f' is a function, 'i' and 'j' are pixel coordinates. Image can be represented as a matrix of integer values (Sheikh & Bovik, 2006).

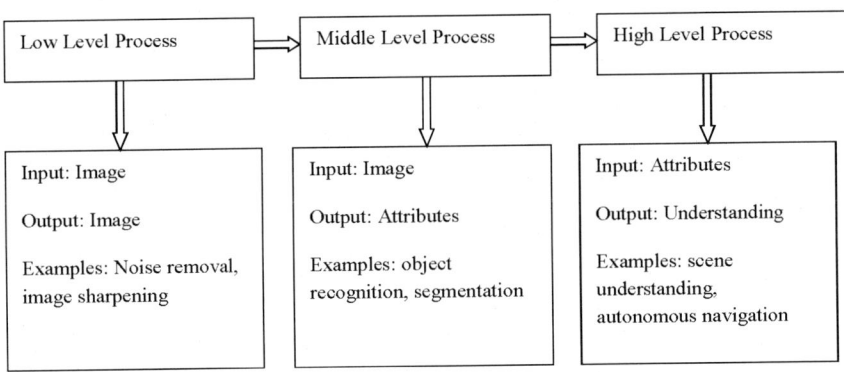

Figure 1. Three level processes of image processing.

The main ideas of image processing are to improve the appearance for human perception and analyze the image for autonomous machine perception. The distinction between image processing and computer vision is tricky in the continuum from image processing to computer vision. This continuum can be broken up into low, middle and high-level processes, as shown in Figure 1.

Low-level processing operates directly on image pixels for minimal information abstraction by performing preprocessing operations on the images. -level processing can be understood and done without prior knowledge of objects in the picture. Mid-level processing operates directly on the image to extract attributes as outputs from the image. The features may be either partitioned images, labeled regions in the image, or classified objects in a snap. High-level processing performs maximum abstraction of information from the image. It performs image analysis by operating on the image's texture, region and objects. The abstraction can be attained efficiently with a priori knowledge about the things in an image.

1.1. Basic Modules in Image Processing

Key modules of image processing are broadly groped into five categories: image formation, image enhancement, image visualization and image analysis and image management, as shown in Figure 2 (Brosnan & Sun, 2004).

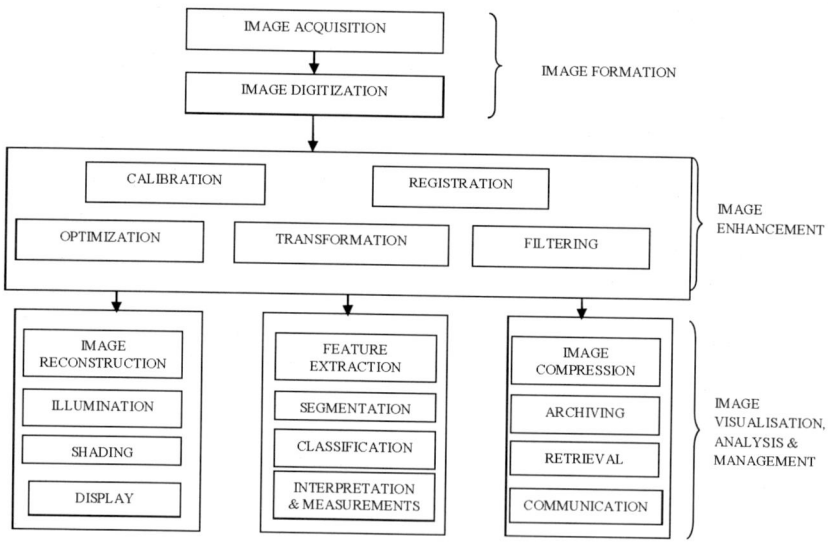

Figure 2. Primary modules of image processing.

1.1.1. Formation

Image formation involves acquiring an image (through any capturing device) and keeping it in a digital form (as an image matrix). A picture, in general, is represented by two-dimensional functions as f(x, y), where f is a function for the coordinates value of x and y. The function for an image must always be nonzero and finite, $0 < f(x, y) < \infty$. The positive scalar value function is determined by illumination from the source and reflection or transmission from the object. The part of an image is in a continuous form and needs to be converted to the digital (discrete) format for analysis. Digitization of pictures is made possible through the concepts of sampling and quantization (Vajda, 1994).

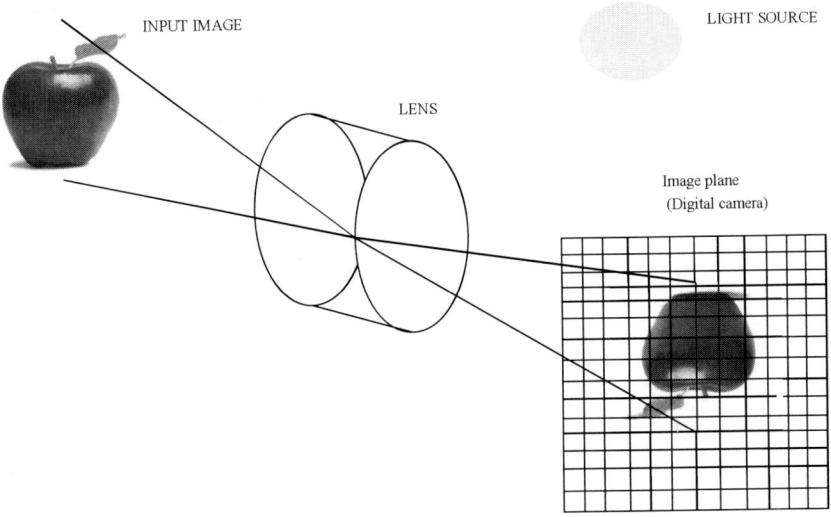

Figure 3. Simple image formation using a digital camera.

Image, in the form of a 2-D array with rows and columns of discrete coordinates. Sampling is the process of digitizing coordinate values in a function of an image. The continuous form of an image is sampled by taking equally spaced samples. The spatial coordinates x and y are converted to discrete coordinates by sampling due to the limited spatial and temporal resolution. After the sampling process, the image is required to be quantized. Quantizing is a process of digitizing amplitude values in a function. These values represent an image's intensity values, which can be made discrete by dividing the intensity scale into discrete intervals ranging from black (0) to white (Ebrahimi et al. 2012). Quantization is performed due to the limited intensity resolution. Function f(x, y) is converted to a digital image by performing the sampling and quantization process in a picture as in Figure 3.

1.1.2. Enhancement

Image enhancement is a manipulation process performed on an image to provide better information to satisfy the user's requirements. It effectively displays or records data for successive visual understanding. This process does not change inbuilt details on data, but it changes the dynamic range of specific features for localization. Most pre- and post-processing algorithms are developed based on these image enhancement techniques. Image enhancement techniques are broadly classified into two categories, spatial and frequency domain techniques as in Figure 4 (Maini & Aggarwal, 2010).

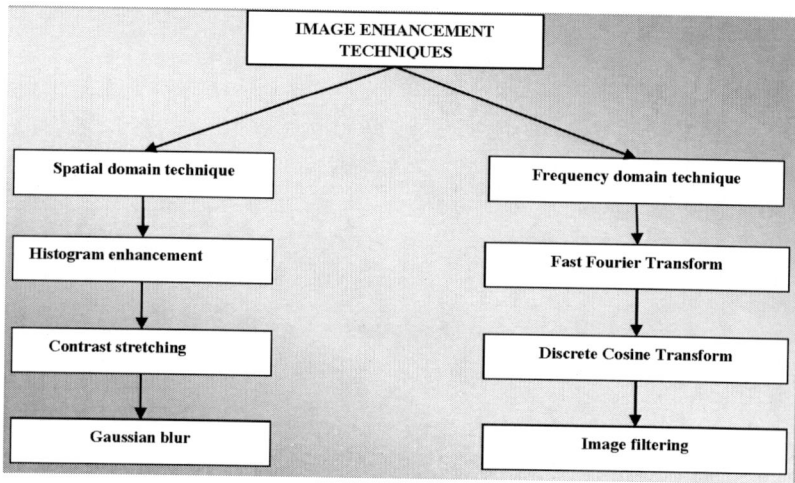

Figure 4. Image enhancement techniques.

Image sharpening and Image smoothing are two essential aspects of image enhancement techniques. In Image sharpening, image information is detailed such that it contains high spatial frequency components of the image. Most techniques contain some form of high pass filtering, which can be done in both the spatial and frequency domain. However, high pass filtering alone can cause the image to lose its contrast. In Image smoothing, image information is reduced because it contains low spatial frequency components of the picture. A larger mask size would give a more significant smoothing effect. Too much smoothing will eventually lead to blurring. In the frequency domain, image smoothing is accomplished using a low-pass filter.

The *spatial domain technique* performs direct manipulation of the image pixels. In this technique, operation (linear or non-linear) is performed based on the neighborhood pixels for the input image, giving enhanced images as output. The neighborhood of domain can be any shape, but generally, rectangular shapes are used (3x3, 5x5, 9x9, etc.). The manipulation can be done based on point processing or neighborhood processing. The gray level transformation or point processing, has a 1x1 neighborhood to manipulate the image using s = T®, where 's' is the enhanced image for the input image 'r' and 'T' is the manipulation performed on the input image. Some fundamental gray level transformations are image negatives (complement), log transformation, power-law transform, piece-wise linear transform, gray level slicing and bit plane slicing. Neighborhood processing, also known as spatial filtering, has a larger neighborhood to perform a predefined operation to enhance the image. This predefined operation is manipulated using spatial mask, filters, kernels, templates, and windows. The filter's character is determined using a spatial mask with a predefined operation. The resultant filtered image is attained when the center point of the filter mask visits each pixel in the input image. The procedure performed on the image pixel is linear; then, the filter is referred to as a linear spatial filter; otherwise, it is referred to as nonlinear spatial filter.

Correlation and convolution are the two essential terms in spatial filtering. Correlation determines the degree of likeness among data sets. Correlation is a process in which the kernel is moved over entire pixels in an image and the predefined manipulation is performed in each pixel position. Image convolution is similar to correlation except the kernel is rotated to 180 degrees before performing manipulations. The image convolution operation is used widely for enhancing effects (like blurring, edge sharpening, etc.) in the photos, as in Figure 5.

(a)　　　　　　　　　　(b)

Figure 5. Median filtering operation performed on the original image (a) to enhance the image as in (b).

The *frequency domain technique* performs indirect manipulation of images by transforming the pictures using convolutions or window kernels. In some instances, the image can be processed better in the frequency domain than the spatial domain. The image is first converted to the transform domain in the frequency domain. The manipulations are performed on the transform domain using forward transformation. After manipulation, it is again restored to the spatial domain by doing the inverse transformation. The resultant image in the spatial domain will be better and improved to satisfy the end user's need. Thus in frequency domain, modification in image position results changes in spatial domain. Alterations in the spatial domain are based on the rate of change in intensity values in an image. There are various transforms available in the frequency domain. Some of them are Discrete Fourier transforms, Fast Fourier transforms, Discrete Cosine transforms, Continuous wavelet transforms, Windowed Fourier transforms, Walsh Hadamard transforms, Haar transforms, Hotelling transform and so on.

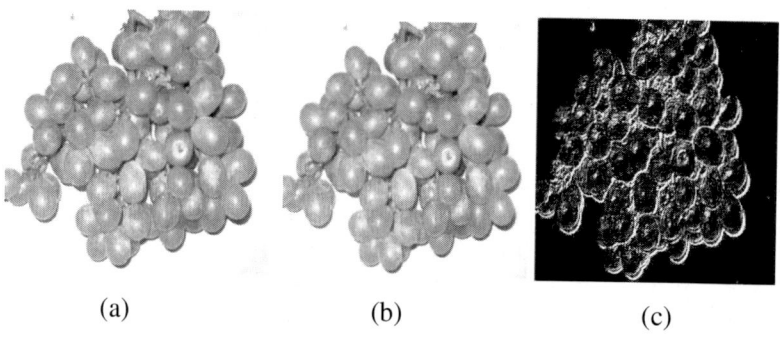

(a)　　　　　　　　(b)　　　　　　　　(c)

Figure 6. Original image (a) is smoothed using image smoothing filter as in '(b)' and is sharpened using image sharpening filter as in '(c)'.

Discrete Fourier transform is one of the critical transform models used in image processing. It represents the image in frequency form. It plays an essential role in the process of image filtering. It is an image processing technique used to smooth, de-blur and restore the image to an enhanced form. The filtering process is performed either as a low-pass filter or a high-pass filter. A low-pass filter reduces the amplitude of high frequencies and has low frequencies unchanged. This filter is very much helpful in smoothing the images. A high pass filter retains high frequencies and soothes the amplitude of low frequencies. This filter is very much helpful in sharpening the images. Theoretically, the spatial domain could be seen easier for smoothing and sharpening operations on the pictures. But in practice, the frequency domain is commonly used for smoothing and sharpening operations on images, as in Figure 6. The filter mask is an operation carried out on the image array set. The image array set has a bunch of scalar components multiplied by the amplitude of the Fourier term. The role of the filtering mask in transformation can be represented mathematically as, $G(m, n) = H(m, n) * F(m, n)$ where F(m,n) is the input image, H(m,n) is the filter mask, and G(m,n) is a resultant enhanced image. The role of the filter mask is significant in noise removal and edge detection. The filter mask helps to make the image understandable, clear and sharper. Different mask models like LoG operator, Roberts operator, Sobel operator, Prewitt operator and so on are used for noise removal and edge detection technique (Shahzad et al. 2009).

1.1.3. Visualization

Image visualization discusses various manipulation methods performed on an image matrix resulting in an improved output image. Image reconstruction is one of the key concepts with in image visualization. In a 2-D image, the resolution of the final output image depends on the co-ordinate positions taken for measurements from the input image on a flat surface. The detector is being used for the measure of the input image. The detector points the direction of the emitter at angle theta concerning the baseline. The detected measurement represents the amount of energy which the object from emitter has not absorbed. The central slice theorem in two dimensions (2D) states that the 1D Fourier transform P(ω) of the projection p(s) of a 2D function f(x, y) is equal to a slice (i.e., a 1D profile). Because the origin of the 2D Fourier transform F(ω_x, ω_y) of that function is parallel to the detector. Image reconstruction is commonly done using the summation method or back projection. The co-ordinate values of input image are used in the back projection method for analysis. The consecutive projections of image are provided with the help of detector at different values of angle theta. Illumination in an image is dependent on the spectrum of scene environment. The visible light of electromagnetic spectrum spans from 400 nm to 700 nm as in Figure 7. Quality of light is dependent on the three basic quantities of light source such as radiance, luminance and brightness. Radiance is defined as the total number of energy that is released by the light source. Luminance can be defined as a measure of the amount of energy perceived from the light source by the observer. Measuring the brightness is impossible and its perception is based on the intensity levels of color in the object observed. The capturing device also has an impact on image display. The magnifying power and light-gathering capability of the lens of a camera determines the quality of the image taken along with the illumination from the light source.

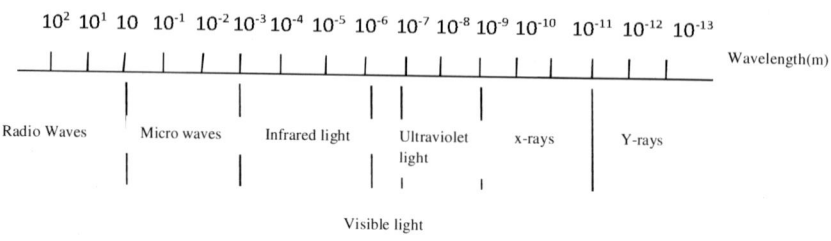

Figure 7. Electromagnetic spectrum.

1.1.4. Analysis

Image analysis is a method used for quantitative measurements, information abstraction and interpretation of data from the image. Image analysis is defined as a process of gathering information from an image. Information gathered from an image can be analyzed statistically using graphs or other statistical measurement tools. Methods like feature extraction, image segmentation, classification, measurements and interpretations come under the image analysis category. The term feature is a continuous collection of pixels representing an image's required object or region of interest. Background is a collection of pixels representing an image's unwanted region of interest. Separating unwanted areas (background) from the region of interest (features) is usually termed image segmentation. Boundary line detection and edge detection also comes under the image segmentation concept. Extraction of features from an image for further measurement or interpretation is termed feature extraction. The binary image is used in the representation of features and background. The background pixels are labeled as 0 (black) and pixels of feature regions are labeled as 1 (white), as in Figure 8. There are various feature extraction and image segmentation methods in image analysis. Some methods are histogram, thresholding, clustering, and so on. The thresholding method sets a threshold value for an image to produce a binary image.

Image classification is the most essential part of digital image analysis. It is a process of classifying all the pixels in an image into one of several predefined classification groups. The two types of image classification are supervised and unsupervised classification. In supervised classification, an image analyst "supervises" the selection of regions representing features that analysts can recognize. In unsupervised classification, image analysis has no role to play because statistical clustering algorithms are used to select classes inherent to data that are more computer-automated. Measurements are made on images after extracting the region of interest from the image. Two principal types of measurements can be performed on an image. The first type performs numerical measures on the position, size, shape, or color of each feature in an image. The other type of measurement measures parameters in an image like total number of pixels in an image, total area, and perimeter.

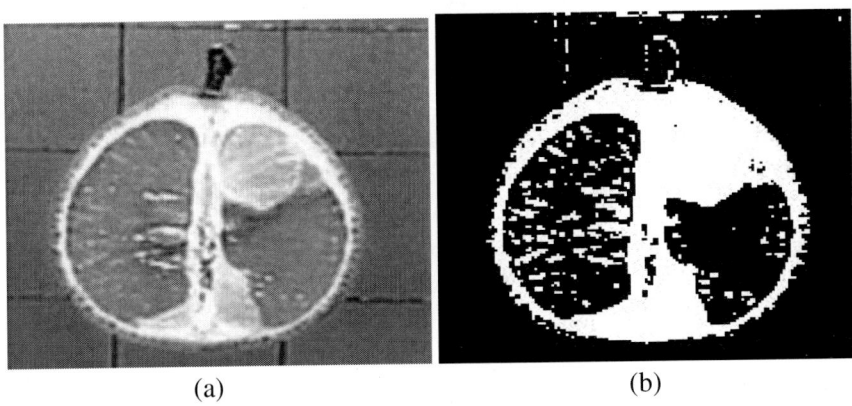

(a) (b)

Figure 8. (a) Original image; (b) thresholded image.

1.1.5. Management

Image management is used in organizing digital images efficiently, providing efficient storage, communication, transmission, archiving, and access (retrieval) of image data. Various techniques dealt with are image compression, archiving, retrieval and communication. Image compression techniques reduce the required data into a compressed or condensed form. Many image compression algorithms are developed based on the concept- the 'Eye is unable to recognize small changes in an image. The compression technique is required to reduce the volume of data to be transmitted (text, fax, images), reduce the bandwidth required for transmission, and reduce storage requirements (speech, audio, video). Two basic key components in compression are reduction in redundancy in digital image data and the properties of human perception, which does not recognize changes in irrelevant data in an image.

Image compression methods are used for digital images and play an important part in digital audio and video size reduction (Bhope & Patil, 2010). Digital audio is a series of signal sample values, whereas the image is a rectangular array of pixel values and video is a sequence of images played out at a specific rate. The samples of neighboring values are correlated with the original sample and a new value is obtained. A compressed version of digital audio, image and video does not represent the exact original information as in Figure 9. There will be some unnoticed reduction or change in data. Perception of the human eye is different for different signal forms. The human eye is less sensitive to higher spatial frequency components than lower frequencies. Two categories of image compression are lossless and lossy compression. Lossless compression is helpful when data redundancies can alone be reduced. Transmission of legal and secure information or document is made possible through lossless compression. Lossy compression is useful when the sensitivity of human perception is possible along with data redundancy reduction. Transmission of digital audio, video and image is made possible through lossy compression, where minor errors are exempted.

(a)   (b)

Figure 9. (a) Original image; (b) compressed image.

Chapter 2

Various Applications of Digital Image Processing

The application of digital image processing is expanding in various fields like medical imaging, aerial and satellite imaging, industrial inspection, law enforcement, defense applications, agriculture sector, etc. Today, each field, either in the private or government sector, has a relationship with image processing modules either directly or indirectly. Their applications in various fields can judge the importance and impact of image processing in today's society. Image processing methods producing higher performance accuracy in one field may have poor performance. Therefore, selecting appropriate modules relating to the application is very important (Dougherty, 2009). Modules can be used either as an individual modules or a combination of modules. The application of image processing in different sectors, as depicted in Figure 10, is discussed briefly in the following section.

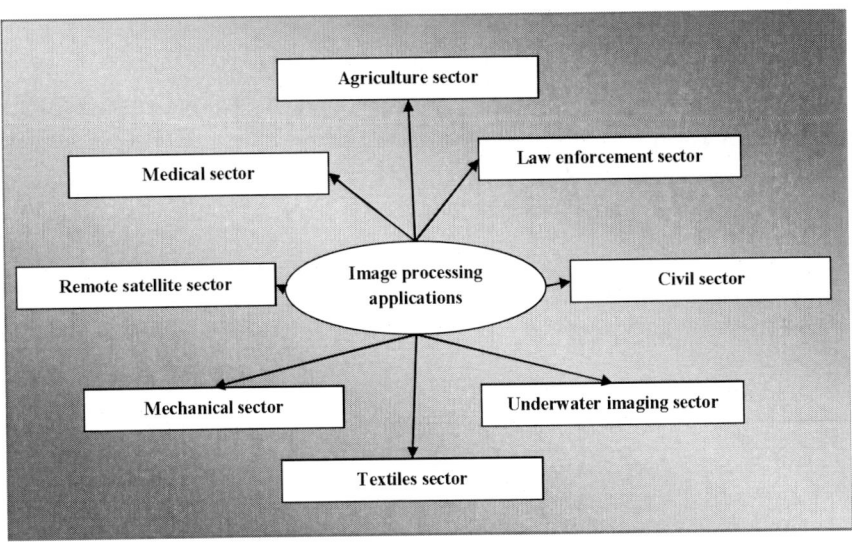

Figure 10. Modules of image processing.

2.1. Medical Sector

Medical image is an interdisciplinary application combining medical physics, applied mathematics and computer science. Medical physics deals with the diagnostic and therapeutic use of electromagnetic radiation, whereas applied mathematics and computer science, are combined as image processing applications. Electromagnetic radiation is used in the medical sector to diagnose diseases in the internal organ of the human body, as in Figure 11. There is rapid development in this sector with computer-aided diagnosis and analysis. This medical image technology has supported doctors for easy diagnosis, surgeries and decision-making.

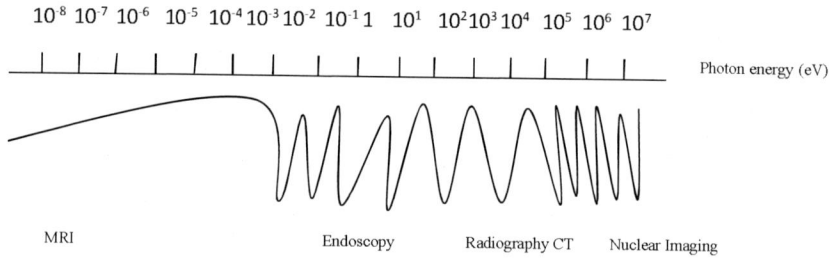

Figure 11. The use of electromagnetic spectrum in different modalities of medical image formation.

Image processing modules related to medical imaging assist in understanding tasks related to disease diagnosis and also it helps to minimize artifacts, reduce distortion, and enhance major required feature. Medical images do not provide an anatomy of a person instead, it provides information about tissue properties like a mass number, density of protons, the mechanical density of the tissue and so on. Medical images are heterogeneous and are related to pictures of tissues and internal organs of the body. The prior knowledge of images is impossible in this field. The utmost care should be taken in the automatic analysis of medical images, for it should not provide wrong measurements. The information and measurements obtained from medical images must have higher reliability and accuracy.

2.2. Law Enforcement Sector

Digital image supports the Law enforcement sector by acting as a powerful source of evidence. The evidential images have a greater value and assist in criminal justice judgment. The evidential image can be either a still image or a moving image. Police officials do the role of management of digital evidential image. The advancement or importance of digital images will soon urge the legislature to form a separate central imaging department in the police department. They have to protect images' originality even from the image generated by officer, other police staff, third parties and the public. These evidential images are subject to entire auditing per legislative requirements before considering them as evidence. The original evidence image is kept secured and only photocopy image is used for any processing and manipulation of images. The processing of evidence image encompasses capturing, recording, storing data and performing various manipulation in the picture. Various image capturing devices like closed circuit television, Body-Worn Video devices, Road safety cameras and so on can be used for gathering evidence in areas where the crime or law disorders occur.

2.3. Satellite Remote Sensing Sector

Remote sensing image processing is an advanced research area that has a significant role in society. It has many real-time applications and most of the activities of daily life are dependent on this sector. Due to its increasing significance, this field has become a multi-disciplinary field. The disciplines like physics, computer science, signal theory and electronics are included. It deals with the collection of information regarding an object or area by analyzing data acquired with a device that is not in contact with the object, area, or phenomenon. The primary purposes served in this sector are monitoring the earth's surface, evaluating geographical, biological and physical variables, analyzing spectral information gathered by satellite and identifying materials on land cover. The impact of remote sensing is more significant when it serves the purpose of urban monitoring, fire detection in dense forests, flood prediction, climatic condition prediction and so on (Sadykhov, 2007).

Various techniques like regression, function approximation, machine learning, classification, restoration, enhancement, feature extraction and so on are implemented in the development of algorithms. Data acquisition depends on energy resources like particles and gases in the atmosphere and the emission of electromagnetic radiation. In remote sensing, sensors make data acquisition possible from any distance (short, medium, or long distance). The sensors can be either *passive or active*. The major energy source for passive sensors is the sun. An example of a passive source is aerial photography. The active sensors have their source of energy. An example of an active source is RADAR. Electromagnetic energy is made used by the majority of remote sensing devices. All wavelengths of electromagnetic radiation are not equally important for remote sensing. The wavelength with different frequencies serves other purposes, as in Table 1.

Table 1. Different applications of remote sensing in different wavelength

Band	Major Application
1	Identifying water areas, discriminating different soil and vegetation areas, and detecting forest types.
2	Measuring the peak green reflectance area of vegetation for vegetation differentiation.
3	Plant species differentiation is made by using chlorophyll absorption.
4	Soil moisture categorization, finding the vegetation type.
5	Distinguishing snow from the clouds, Moisture content analysis
6	Thermal mapping applications, soil moisture distinction
7	Mineral and rock types discrimination
Pan	Multi-spectral image sharpening

The major steps involved in remote sensing image processing techniques are image acquisition, image preprocessing, image restoration, image analysis, image compression and finally, the processed output data. In image preprocessing, geometric and radiometric corrections are performed for data accuracy and image enhancement is performed to improve data quality. The image restoration method is mainly used to serve the purpose of reducing noise in an image, and the possibility of noise in an image is more in this imaging technique. The image analysis is performed for the data's feature extraction, classification, clustering, and segmentation. Image compression is compulsorily performed in satellite images, for they collect and store numerous amounts of data regularly and continuously. The final output image or data is used to assist scientists in decision-making with the information gathered (Thomas, 2009).

2.4. Civil Engineering Sector

Today, the infrastructure sector is developing rapidly with advanced technology. Existing old infrastructures are also modified to the latest advancement. Valuable infrastructure assets are to be maintained and assessments must be made periodically. In recent years, the role of image processing has accelerated acceleration in civil engineering. Innovations and research in this sector have more importance due to automated technologies demand the inspection. The image processing technique is used for various purposes like checking pavement conditions, underground pipeline inspection, steel bridge coating, concrete surface defect assessment, and so on. Manual review is -consuming, depends on individual ability and experience, the result may differ from person to person, and the probability of error is higher in this assessment. Automated infrastructure assessment is preferred to the manual inspection method due to its accuracy, objectivity, speed, and consistency. In steel bridge coating assessment, the bridge's rust can be recognized and defected regions can be easily detected. An automated system assists managers of infrastructure maintenance in making decisions on coating. The managers decide whether to repaint the steel bridge again or not. The success of steel bridge assessment depends on the efficient evaluation of coating conditions (Gokhale & Yang, 1999).

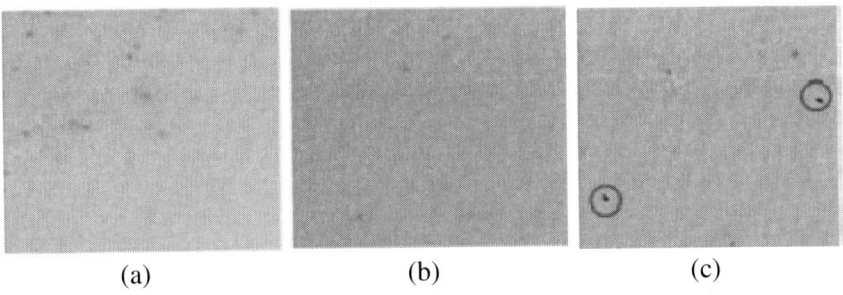

Figure 12. (a) Rust image on the steel bridge. (b) Low contrast image of '(a)'. (c) Identification of noise from '(b)'.

The concrete surface assessment identifies surface defects like cracks, air pockets and discoloration. Concrete surface quality is determined by the concrete mixture used for construction. The quality of concrete mixture used for construction determines strength, appearance and durability. The quality standard is fixed for concrete and surfaces of the infrastructure are inspected regularly. If the quality of standard is not followed properly, it may soon lead to visible defects on the surface. This might reduce the life cycle of concrete, and the quality of final construction will also be poor. Proper care can be taken at the time of concrete mixture. The concrete mix can also be assessed using high-resolution 2-D and 3-D images with microstructure information of the concrete. X-ray CT is a non-destructive imaging technique used in concrete image analysis. X-ray CT technology is used to identify the characteristics of different material goods used in the mixture. An automated system on concrete surfaces captures surface images from a predetermined distance from the surface. Primary image processing techniques involved in this application are image enhancement like contrast stretching and noise removal, thresholding, edge detection and image segmentation. The captured image is preprocessed such that unwanted features are removed from the image. Required features alone can be segmented and enhanced to automatically detect a number of cracks, air pockets and amount of discoloration, as in Figure 12.

2.5. Mechanical Engineering Sector

Non-destructive method of analysis and higher accuracy in the automated image processing system has attracted the mechanical engineering sector. The product requirement can be automatically assessed through which production rate could be decreased by reducing scrap components. Defects and other flaws in a product can be easily identified by manufacturing companies in real-time with higher reliability and lower expense, as in Figure 13. The manufacturing company's duty is to produce products with minimized cost and high quality and inspect the designed product for guarantee assurance. Manual inspection is time-consuming and it requires prior knowledge and experience. Inspection result is prone to error and result might differ from person to person. The advantages of automated systems have attracted the manufacturing companies and engineers for implementing them in their regular activities. Manufacturing industries of thin film materials like semiconductor wafers, photovoltaic and photoconductors, plastics, graphic art papers and films, optical films and so on have fully automated their inspection process. Other sectors are motor manufacturing companies (cars, two wheelers), durable product manufacturing companies, aircraft assembly, and so on. Surface defects on a product's steel sheet are identified and the defect's location is marked for further analysis (Zelelew et al. 2008).

Methods like an artificial neural network, fuzzy expert system, condition-based reasoning, random forest, wavelets, support vector machines, content-based imaging techniques, and so on are used to monitor the machine condition and diagnose the defect. Content-Based Imaging technique is a popular methods used ithe vision-based analysis. In this technique, features of the data set are extracted, and data dimensionality is reduced, but the elements of the data set are retained. Statistical characterization of linear or non-linear transforms is combined to extract features from the data set. This technique predefines features in the database so that the execution can be done with a similar classification algorithm. This imaging technique is mostly applied in detecting surface defects on the product along with their location. Support vector is a rarely used technique in monitoring and identifying defects in machines. The performance of using this technique has a higher degree of accuracy in the feature classification (Lee, 2011).

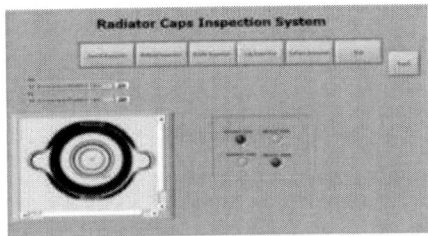
Radiator Cap Inspection

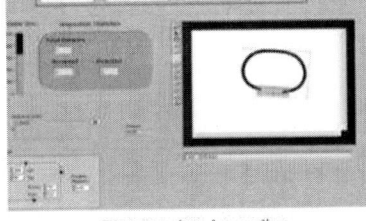
Ring travelers Inspection

Figure 13. Automatic inspection of the radiator cap and ring traveler.

Wavelets are also used in machine defect analysis with time-frequency signals. Signals from an image are obtained and a simple FFT transformation is applied to extract the principal feature of signals -based maintenance is another common approach used for a machine or product maintenance. In this method, the product is monitored and information's is gathered regularly. The maintenance work is done only when there is an indication from monitoring data. The use of this technique can avoid unnecessary maintenance work. Methods like image thresholding, image segmentation, image enhancement, edge detection, artificial neural network and so on are applied for feature extraction. Acquired images can be enhanced by image preprocessing techniques like image thresholding and enhancement. Image segmentation and edge detection are also used to extract features from the image for further analysis.

2.6. Underwater Imaging Sector

The exploration underwater has accelerating importance with multi-disciplinary exposure in various fields like physics, biology, geology, archaeology, etc. Acoustic vision is a recently accelerating field in underwater images and is used to describe the processing of images. Methods used in acoustic vision are image acquisition, segmentation, reconstruction and object recognition. Submerged objects, underwater living things, plant species, etc., can be easily identified. Image acquisition is dependent on sensors rather than signal frequencies. Optical sensors collect fine resolution data but has the disadvantage of limited range in acquiring data and lack of 3-D information (Arnold-Bos et al. 2005).

The acoustic imaging system is a common method used for acquiring images. Acoustic imaging deals with acquiring and processing images using an acoustic system. Acoustic imaging is used in various sectors like medical imaging, aerial imaging, and underwater imaging, etc. Data with small and large visibility can be easily identified and images can also be captured but with poor quality and coarse resolution. Images obtained through this technique are ambiguous and unclear. The factors responsible for the degraded image are structure, less environment, light reflection, absorption and scattering. The reflection of light depends on the density of sea; only a part of the light enters water in depth with reduced color wavelength. Some other particles and molecules absorb a certain amount of light. These factors make the image darker and noisy. Recent scientific development has made it possible to capture and improve images of underwater scenes. Technologies like autonomous water vehicles, visually guided swimming robots, sensors and optical cameras are used in capturing underwater images with recent innovations (Bazeille et al. 2006).

Image processing techniques involved help to rectify the incorrect illumination, reduce noise, and enhance the image by adjusting colors. The commonly used image processing techniques are image restoration and image enhancement methods. Image restoration reduces coarse and noise in an image (Murino & Trucco, 2000). Convolution masks and filters like low pass filter, linear smoothing filter and median filter are used for this purpose. Image enhancement is a simple method used to enhance the image with better visibility, as in Figure 14. Contrast stretching, Gaussian-blur, Log-Gabor and Histogram equalization are some basic image enhancement methods implemented in underwater imaging. Contrast stretching is used to improve color contrast and histogram equalization is used to equalize color contrast in an image. Edge detection methods are also used to detect edges in an image and to reduce noise. Color space conversion is also performed to increase the original color in an image which is affected due to the lighting factor. Image is converted to HSI or YCbCr color space to perform all operations on the individual color channel. The stretching of saturation and intensity value in an HSI color space helps to improve an image's color. Luminance channel (Y) in YCbCr color space enhances the brightness of the original dark image. Manipulations in different color spaces help to speed up the analysis process (Iqbal et al. 2007).

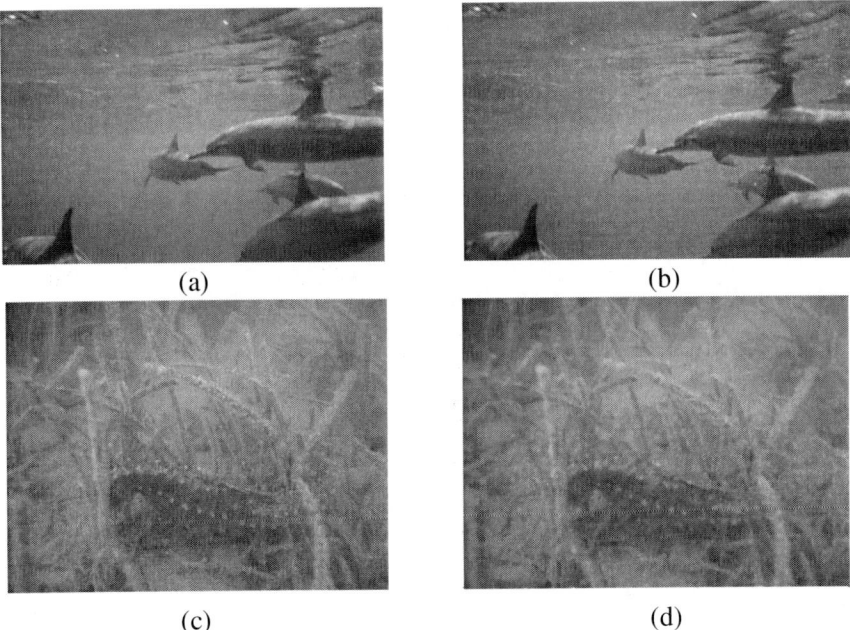

Figure 14. (a) Underwater image of a dolphin. (b) Enhanced image of '(a)' using filtering method. (c) Underwater image of a seahorse. (d) Enhanced image of '(c)' using filtering method.

2.7. Textile Industries

To overcome the disadvantages of off loom inspections in the Textile industry, computer vision system is very useful in identifying defects in loom inspections of fabrics. The fabric quality is graded as first and second quality. The first quality is free from total damages, including major and minor defects in the fabric surfaces. The second quality suffers from major and minor imperfections in the fabric surfaces. In recent years, production must be speed enough to meet the tough competition and the demand of the customers must be satisfied in time. So, the detection or identification of defects in fabrics manually consumes more time and creates chances to delay the production process and distribution to customers. Inspection individuals also have opportunities to make an error in inspection (Bhope & Patil, 2010). Online fabric inspection has been introduced in textile industry to overcome the disadvantage of offline fabric inspection. Image processing modules have been used to identify the fabric's structural defects.

Image processing modules like image acquisition, image enhancement methods like fast fourier transforms, histogram equalizations, and spatial fourier analysis was applied in the online automated fabric analysis system. The input image is the fabric material. As computer vision is a non-destructive method, there is no chance of any damage to the fabric during the inspection process. Methods like Histogram equalization have been used to enhance the image quality for further analysis. An enhanced image produces a better result than the normal acquired image. Fast Fourier Transforms have been applied to analyze the fabric quality. Image processing application is thus spreading over various sectors and has become an integral part of those sectors by fulfilling their needs.

Chapter 3

Role of Image Processing in the Agriculture Sector

Agriculture is an important sector that provides basic needs and food for human beings. The sector concentrates on producing qualitative fruits, vegetables, cereals, pulses, etc., with minimized expenses. Advances in science and technology have made a new revolution in the agriculture sector. The agriculture sector is integrated with information technology to increase its standard by using the automated system in various activities. New technologies like GPS technology, precision agriculture, robotics, and sensor networks have emerged with recent inventions and innovations. The solution for multiple problems has been easily identified with higher speed and accuracy (Butz et al. 2005).

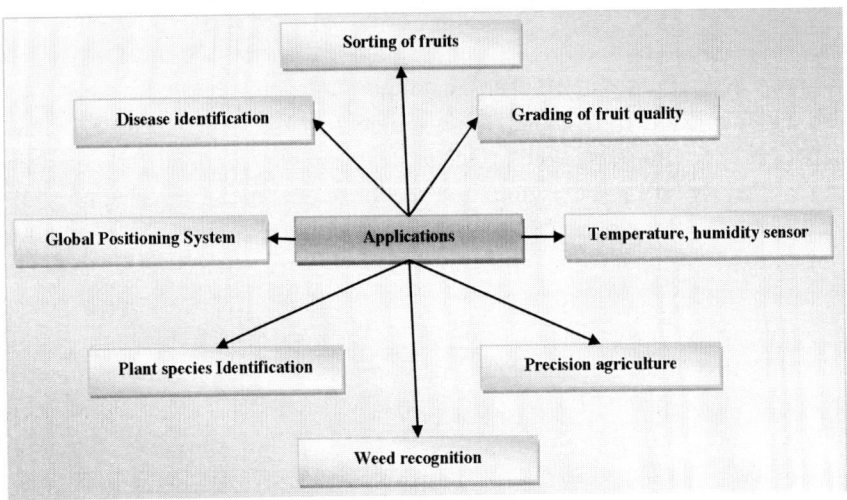

Figure 15. Applications of image processing in agriculture.

The automated system has been developed to solve various problems faced by farmers, as in Figure 15. It eases the work of farmers and assists them in maximizing their production with minimum cost. The work of human labor is time-consuming, individual decisions based on assumptions and expensive. Human labor is being reduced to a certain extent with the replacement of automated systems. Computer vision, machine vision and image processing are various techniques developed in the automated system to serve different purposes (Bhattacharyya, 2011). The results of these techniques are accurate, reliable and consistent.

Various sensors are used in the agriculture sector for monitoring temperature, humidity, land wetness, leaf wetness and climatic conditions. The proper use of these sensors could predict climatic changes in the near future, which would direct farmers to a proper plan to yield good growth of crops. The essential requirement of a farmer is to have adequate monitoring of crops. This automated monitoring enables supervision of plant diseases, nutrient deficiency, weed growth, and maturity of crop. Weed is an important threat to farmers as it reduces crop production and quality. Therefore, more attention is needed to monitor the weeds. The use of herbicide is one of the common methods used to control the growth of weeds. With the latest innovations, *weed recognition* is automated such that the system automatically distinguishes the weeds from the crops. An automated system monitors the crop regularly and destroys the weed. The techniques used in the weed recognition are fuzzy logic, neural network, and color image processing.

Precision agriculture is used in farm management to observe, assess and control agricultural areas. The main aim of precision agriculture is to maximize profit with minimum input and optimal use of the resource. Farmers are to be well-informed about the technology and its workings. Proper training is required for farmers to acquire information about precision agriculture. GPS and GIS are the technologies used in agriculture equipment in precision agriculture. The geographic information system is used to identify all available data and Global Positioning System is used to determine the position of objects on the globe using satellite signals. An important application is the identification of diseases in agricultural crops. Reasons for *disease infections* is climate change, fungi, bacteria, viruses and health deficiency. Farmers can protect the crops from damage through an earlier stages of diagnosis. Techniques like support vector machine, fuzzy logic, neural networks, image segmentation and image enhancement are used in an automated systems to diagnose plant disease. Disease type can also be easily identified and action can be taken by farmers to control the disease. *Plant species identification* is also a critical application beneficial for botanist. Content-based image retrieval is used in species identification from the collection of species images. *Automatic sorting and grading* agricultural produce is a post-harvest application of image processing. The industrial sector is benefited from automated grading and sorting of fruits and vegetables by applying a Non-destructive method. It is more advantageous than human labor with more accuracy, reliability, speed and consistency. Fruits with different maturity stages are easily identified and separated. The damaged fruits and vegetables can also be easily identified and removed for marketing. This automation helps consumers to buy qualitative fruits and vegetables from the market. These applications of image processing in the agriculture sector are discussed briefly in this section.

3.1. Techniques Used in Plant Disease Identification

The emergence of new pests and plant diseases are major threats farmers face in recent days. Unpredictable climatic conditions and pioneer changes in society are major sources of emergence of new diseases and pests. The most common sources of disease infections on crops are the pathogens like bacteria, viruses and fungi. The speed of disease-spread in crops depends on the resistance capability of crops. The infection in plants gradually reduces resistance and increases susceptibility. Symptoms of the infection can be noticed in crops starting from their earlier stage. Each crop has different behavior due to disease. Symptoms of disease vary among crops. Common symptoms can be mentioned as a change in color, shape and size of leaves, pre-mature falling of fruits, the appearance of dots or spots in fruit, stem and leaves, bulging in roots, or appearance of knots in the roots. Certain symptoms are directly visible by the naked eye and certain symptoms that cannot be visibly viewed. In the traditional method, diseases were diagnosed by a subject matter specialists. The advancement and higher accuracy of disease diagnosis in the medical sector paved the way for image processing in the agricultural industry for disease diagnosis. Various image processing methods are used for disease diagnosis and damage percentage measurement in the crop. Image acquisition is a common technique for acquiring the image of an infected plant or specified region. Region-based segmentation algorithm has also a higher impact in identifying the damaged regions without considering any illumination conditions for image acquisition. There must be a standard specification for acquiring an image and the same standard must be followed throughout the acquisition process. Preprocessing, analysis, and post-processing methods vary from crop to crop according to suitability (Patil &

Raj Kumar, 2011). Various possible diseases in crops and the image processing techniques used for diagnosis are discussed briefly in this section.

3.1.1. Pomegranate Plant Disease

Image processing technique has been developed for the automatic identification of diseases and the degree of damage in the pomegranate plants. The major possible diseases in pomegranate are anthracnose, bacterial blight and wilt complex. Anthracnose disease develops dark black spots in pomegranates as in the Figure 16(a). Soft computing methods and machine learning techniques were combined in the automatic detection of disease in the pomegranate plants. Image processing modules like image acquisition, image resizing, filtering, segmentation and morphological operations were implemented in this algorithm. There are various machine learning approaches like decision tree learning, association rule learning, genetic programming, inductive logic programming, artificial neural networks, Bayesian networks, clustering, representation learning, reinforcement learning and support vector machines applied in the algorithm for disease diagnosis. Clustering methods are used in the identification of a diseased regions in the fruit, as in Figure 16(b). Machine learning techniques support vector machines for classification, regression and feature selection. The soft computing method fuzzy logic is used for determining uncertainty and haziness in the development of the automated systems (Sannakki et al. 2011).

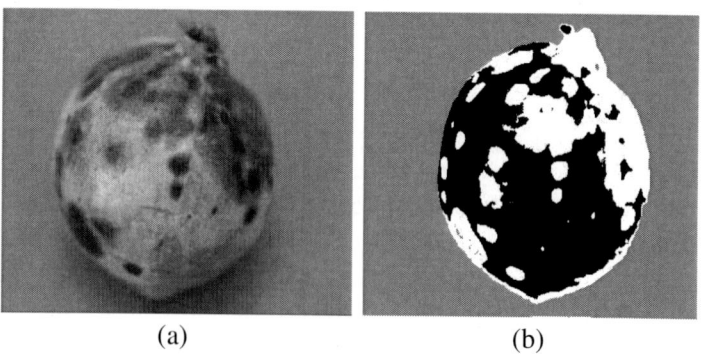

(a)   (b)

Figure 16. (a) Anthracnose disease in pomegranate fruit. (b) Clustering method used in the diagnosis of anthracnose disease in pomegranate fruit.

3.1.2. Citrus Plant Disease

Disease diagnosis of citrus greening and citrus canker in citrus trees can be made more accessible through the use of image processing methods. Citrus greening reduces fruit quality, production and even leads to tree death. Symptoms of this disease are usually found on the leaves of the affected tree. Tree leaves with nutritional deficiencies like iron, manganese, zinc, and nitrogen is easily prone to this infection. Symptoms are different on leaves for each deficiency. Citrus canker-infected fruits have dark brown spots, as in Figure 17(a), by reducing the nutrition values in fruit. Diagnosis of this disease helps to reduce the rapid spread of the disease and avoids lower productivity of fruits in the tree. The algorithm developed for automatic diagnosis uses the color co-occurrence method in image analysis (Blasco et al. 2007). The digital microscope was used for acquiring color citrus leaf images. The acquired RGB image was converted into the HSI color space model for manipulation in the H component, as in Figure 17(b). A color co-occurrence matrix was generated for three components of the HSI color space. Texture features were analyzed using a color co-occurrence matrix (Kim et al. 2009; Pydipati et al. 2006).

(a)   (b)

Figure 17. (a) Citrus canker disease infection in orange fruit. (b) H component of the HSV color space transformation for easy diagnosis of the disease.

3.1.3. Tomato Plant Disease

Tomatoes are a common vegetables produced and consumed throughout the world. The infection of disease lowers the production of crops and even kills the crop. Various possible diseases in the tomato plants are septoria leaf spot, anthracnose, early blight, late blight, *Fusarium* wilt, *Verticillium* wilt, bacterial spot, bacterial speck and viruses. Fungus *Colletotrichum* causes anthracnose and it mainly attacks the fruit of tomatoes as in Figure 18(a). The symptoms are small, circular, indented spots on the fruit's skin. Fungus Alternaria solani causes early blight and it mainly attacks the leaves of the plant. Black or brown spots appear on old leaves with dark edges. Fungus Phytophthora infectants cause late blight and it infects the young and old leaves. It appears as water-soaked areas with irregular, greenish-black blotches. Earlier diagnosis of infection in tomato leaves helps to identify the degree of damage to crop and also assists in recover from the infection and reduces the severity. Color-based analysis of image was performed for diagnosis of infection as in Figure 18(b). The image was separated into red, green and blue color components individually. Color moments were used to calculate the color value distribution in the image in each color component. Nine features extracted were used for disease diagnosis (Monavar et al. 2011).

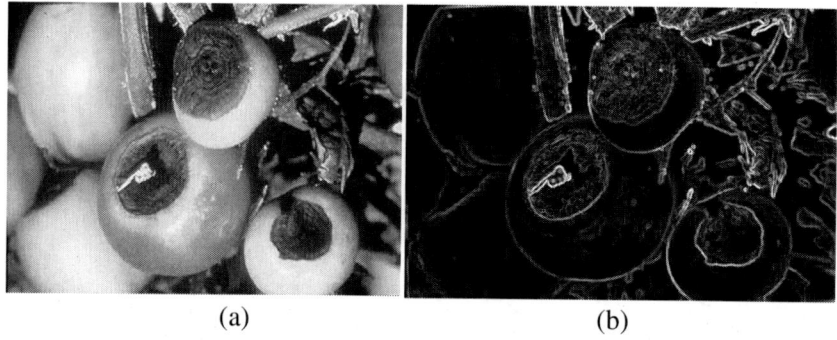

(a) (b)

Figure 18. (a) Anthracnose disease infected tomato. (b) Color analysis performed on the infected tomato.

3.2. Precision Farming

The development of modern information technology and agriculture science tools has made it possible to merge these sectors, leading to the rise of precision farming. Precision farming is an information and technology-based farm management system to identify, analyze and manage variability within fields for optimum profitability, sustainability and production of land resources. Precision farming can also be defined as an application based on soil, weather and crop requirements inputs to maximize production, profit and quality. Precision farming is beneficial to farmers as input prices are increasing and commodity prices are decreasing nowadays. Precision farming is considered a better alternative to traditional farming regarding profit and production. This technological precision farming can assist farmers in better decisions regarding optimal crop production. Precision farming involves proper understanding and efficient use of natural resources found within a field.

Precision farming is also referred to as Global Positioning System agriculture or variable-rate farming. Devices used in precision farming play an important role, and the maximum utilization of technology yields higher profit returns for farmers. Precision farming differs from traditional agriculture by managing fields as small areas and not as a single field unit. Economic and environmental benefits must also be considered along with field management. The barriers to widespread adoption of this technology and the possible benefits are considered major related issues in precision farming, as in Figure 19.

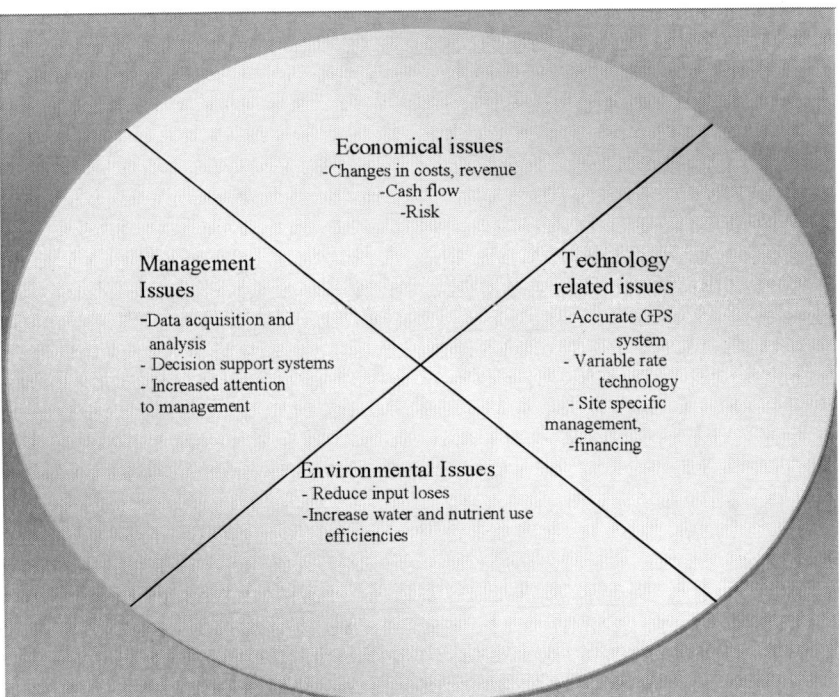

Figure 19. Critical issues related to precision farming.

3.2.1. Technological Tools Used in Precision Farming

Various technological tools are available to collect and implement information effectively. The success of these technologies depends on proper understanding by farmers. Farmers must take appropriate care to understand the technologies and implement them. In general, technological tools include hardware, software and best management practices. The specialized tools used in precision farming are depicted in Figure 20. *Global Positioning System (GPS) receivers* identify their location by receiving signals broadcasted by Global Positioning System satellites. The information received is –ti6me about the location that assists in mapping soil and crop measurements. GPS receivers are mounted on implements in the field, or it can be carried to the field. The uncorrected GPS signals are compared with land-based or satellite-based signals to avail the position correction called a differential correction. The type of differential correction and its coverage capabilities must be considered while purchasing a GPS receiver (Staford, 2000).

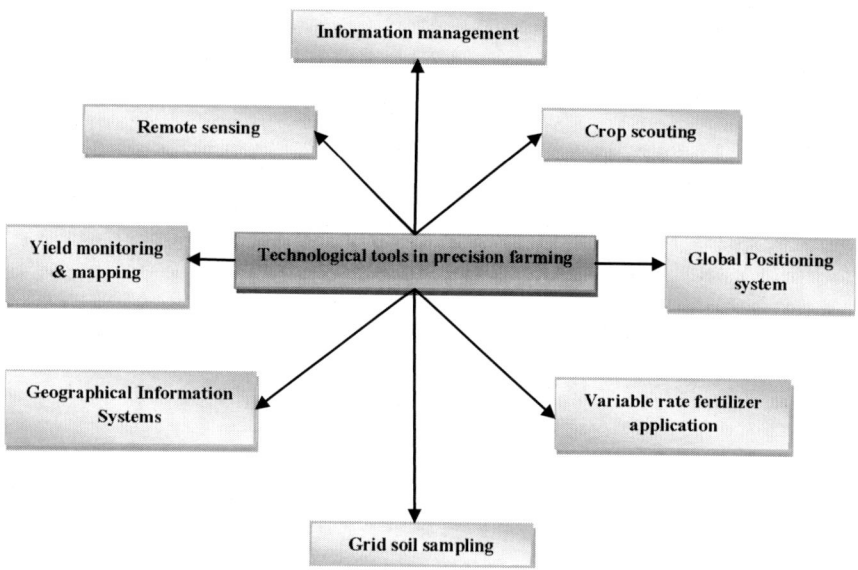

Figure 20. Technological tools used in the precision farming.

The *yield monitors* measure and record grain flow in the grain elevator. These yield monitors have connected with a GPS receiver to avail the required information for *yield maps*. The yield map assists farmers in deciding the input requirements in the field, such as irrigation level, pesticides, fertilizers, etc.

Grid soil sampling is similar to regular soil sampling about principles but varies in intensity of sampling. Grid soil sampling generates an application map providing information about the nutrient requirements. This application map is given as input to the computer mounted on a *variable rate fertilizer spreader*, which supports determining the amount and kind of fertilizers required for soil.

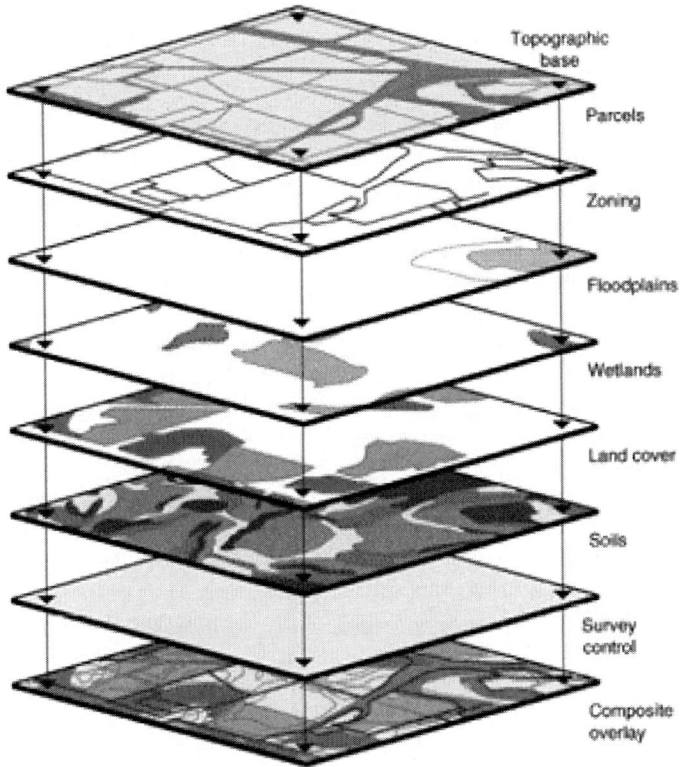

Figure 21. Data layers of map produced by GIS for analysis of information.

Crop scouting is used for accurate analysis of crop yielding capability, weed area in crop and quantity of infection due to insects and climatic conditions. Crop scouting is apart of pest management and part of pest management for farmers. The nutrient state of the crop can also be determined using crop scouting. The location of flooded and eroded area can be observed through crop scouting to take necessary recovery steps. *Geographical information systems* (GIS) are used to generate maps, as in Figure 21, with location information and other feature characteristics about the location of the geographical area. The visual perception of geographical information in the map can be used for interpretation. GIS provides information through various data layers of stored data maps. Agriculture GIS is used to gather, analyze and store to gather, analyze and store information related to soil survey map, yield map, crop scouting reports, and nutrient levels in the soil. GIS data layers help in the evaluation and manipulation of retrieved data to take decisions. GIS is very beneficial to the agricultural sector of its capability to visualize and analyze the agricultural environment. GIS plays an essential part in determining the profit for farmers by balancing the farm inputs and outputs.

Information Management is an important tool in precision farming used for advancing managerial skills along with the information database that provides continuous information. Recordkeeping is an integral part of GIS for storing all acquired maps at different times. A large number of the map with information collected by GIS is to be stored for further analysis and decision-making. The information database used for storing these maps must be persistent for continual access to maps and must have high storage capability. Information management can be of optimum use for farmers when given special education for the capable management and use of technology.

3.3. Crop and Land Assessment Using Remote Sensing

Remote sensing is an important critical data source in GIS for accessing data acquired through satellites. It is a process of gathering information about an object, area, or occurrence without being in contact with it. Data sensors used in remote sensing are hand-held devices mounted on satellites, manned or unmanned aircraft, or directly on farm equipment. A passive system is the most commonly used remote sensing type to sense electromagnetic energy reflected from plants. The factors considered important in remote sensing are reflectance of visible light energy from external sources. The external source of energy for passive systems is, in general, the sun. Information gathered through satellites has been increased through the use of image sensors. Remotely acquired images through satellites can be accessed and analyzed in their digital form. The advancement in image processing techniques like image enhancement, restoration and analysis has made remote sensing progress independently in advance of GIS. The image processing software package is combined with remote sensing for better analysis and interpretation of gathered data. The main aim of remote sensing is to monitor the earth's surface and thereby measure geographical, biological and physical variables to identify the materials on the land cover for further analysis, as in Figure 22.

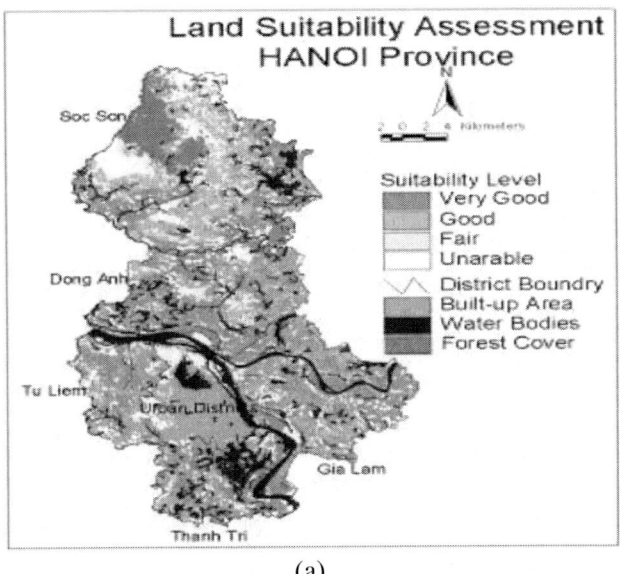

(a)

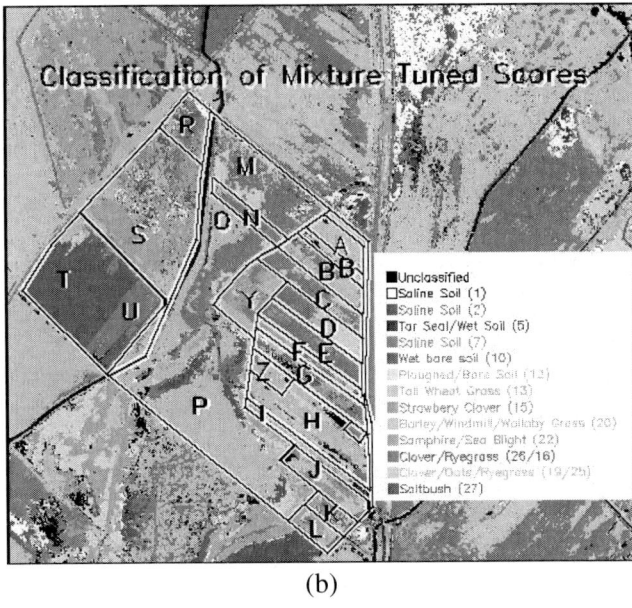

(b)

Figure 22. (a) Remote sensing used in land suitability assessment. (b) Remote sensing used in the scoring of the classification of soil.

3.4. Plant Species Identification

Plants are identified by their morphological characteristics like texture, size, shape and color of the leaves and flowers. The morphological features of plants are observed manually to determine the plant species. Plant leaves are considered a major feature for identifying plant species. Leaves' shape, size and texture are considered in differentiating the plant species. The trained and experienced botanists' support is needed in species recognition. It is a really tiresome job for consumers and traders to decide on plant species. To overcome the difficulty of species recognition in plants, the latest innovations in information technology like smartphones with the android application, real-time image capturing devices with laptops, palmtops, etc., can be introduced to solve the problem domain (Femat-Diaz et al. 2011). Databases were used to store images of various species of plant leaves as in Figure 23.

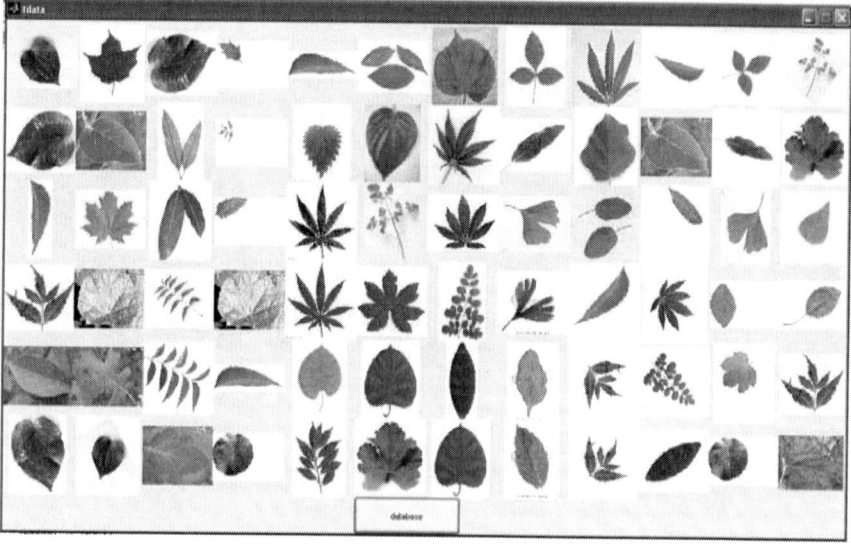

Figure 23. Image database of the leaves.

Leaf classification was performed using image processing methods in actual time applications. Image processing application is available on most mobile platforms. These latest technologies are user-friendly and helps in capturing and automatically identifying the species name in a fraction of second as in Figure 24 (Abdul Kadir et al. 2011; Yang & Ramaswamy, 2000).

The public benefited, for it assists them in acquiring knowledge about species in their environment.

Figure 24. Android based application in the plant species recognition.

Shape analysis on leaves is done based on the leaf boundary. There are two basic leaf analysis approaches: region-based and boundary-based (Bama et al. 2011). Moment descriptors are used in region-based classification approach, and object contour is used in the boundary based classification approach. Consideration of shape characteristics of leaves alone may lead to false classification because there may be two or more species having similar leaf shape. Therefore other features must be considered in classification for higher reliability in the automated system. Other consideration features are the color and texture of the leaf (Patil et al. 2011; Zheng & Wang, 2010).

Color can be considered as the second primary feature in classification, and result accuracy can be greater when color feature is combined with shape feature. The classification efficiency depends on the shape representation used and the technique implemented for shape matching. Recognition of objects was made easier using invariant moments (Murakami et al. 2005). The support vector machine, artificial neural network, polar Fourier transform, image correlogram and general regression methods were used in the image database's pattern recognition of input images. Image correlograms were used for information retrieval from an image. Parameters calculated are aspect ratio, leaf diameter, leaf dent, leaf vein, gray level occurrence matrix, color moments and the invariant moments. Use of these parameters in the measurement of leaf image assists in the analysis of leaf (Sakamoto et al. 2012).

3.5. Fruit Sorting and Classification

Fruit markets nowadays have more responsibility to distribute a variety of fruits. The demand for fruit classification has been increased as different varieties of fruits come to market in large quantities and distributed immediately to various retail shops. Manual is time-consuming and a tedious, repetitive job to classify tons of fruits in a shorter time. Hence there is a need for -based fruit classification which also assists in packaging fruits by using optimal packaging configuration. It also reduces labor and packaging expenses. Variety of fruits also provides benefits in quality evaluation and defect finding. Fruit image can be expressed by image analysis in terms of size, color, shape, defects and abnormalities. Color and shape are primary properties of fruit images that help better classify. The accuracy based on these two criteria will not be significant because there is more than one fruit with similar color and shape (Seng & Mirisaee, 2009). Hence, the size of an image is also considered as another criterion to improve the accuracy. Uniformity in the classification of fruit is determined based on color, size and shape. Classification of fruits can be based either one of these aspects or on a combination of them. Research in this area indicates the feasibility of using these techniques to improve the fruits market standard.

3.5.1. Methods Used for Fruit Classification

There are various phases in image processing that are being used for the classification of fruits. Color image processing and image segmentation are the two methods used for fruit classification based on color (Dadwal & Banga, 2012; Rocha et al. 2010). The regional descriptor is the method used for classification based on size (Gao et al. 2010). Boundary descriptor and feature extraction are used for the classification based on shape (Costa et al. 2011). These methods can be combined together to classify fruit based on three physical attributes, size, shape and size (Bato et al. 2000). These image-processing techniques are discussed in detail in this section.

3.5.1.1. Color Image Processing

Color is a common descriptor for object identification and feature extraction from an image. In color image processing, colors in the image are processed based on their color model (Pass, 1997). Images captured using the cameras are in RGB color model. Three RGB color components are calculated individually using a histogram as in Figure 25 to calculate the mean value of each color component. The mean value obtained from color components sets the threshold value for fruit selection. In some instances, the RGB color is converted to an HSI color model for extracting intensity information from the pixels of an image (Gay et al. 2002). In fruit classification, color coherence vectors (CCV) are being used to compare the images with trained data.

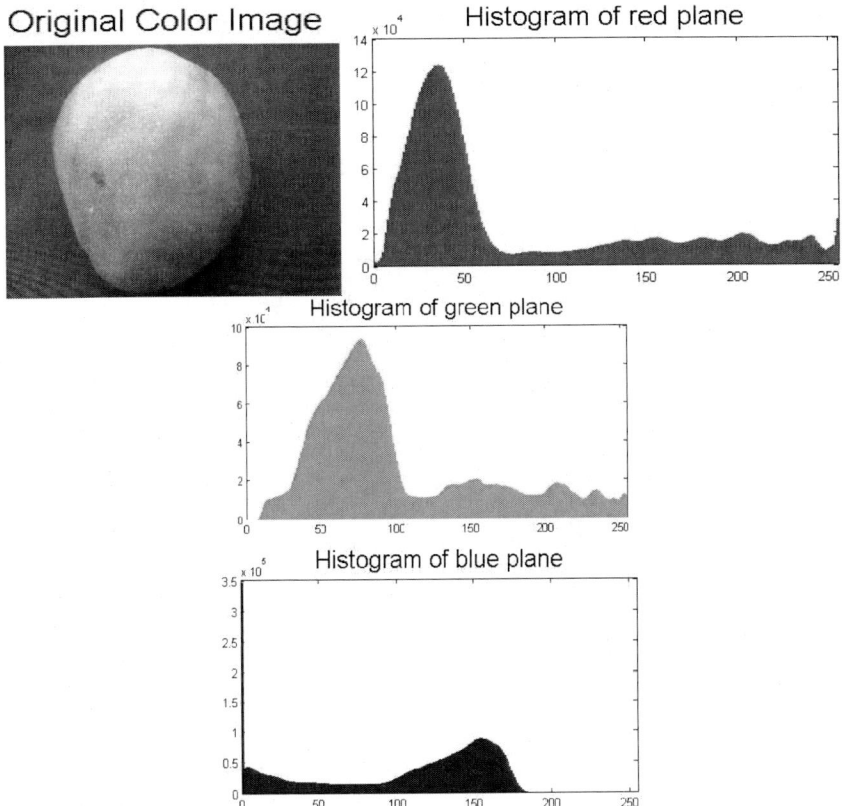

Figure 25. Histogram of the each color component is measured to calculate the threshold value.

Color coherence is defined as a degree where particular pixel color belongs to a larger similar color region. Coherent pixels belong to an image's continuous region, and incoherent pixels do not belong to the region. The CCV blurs and digitizes color space to remove small variations between two neighboring pixels.

3.5.1.2. Image Segmentation

Image segmentation distinguishes the region of interest from the unwanted areas in an image. Thresholding is a standard method used for removing background from an image. The global threshold is used to calculate the binary mask, as in Figure 26 of fruit, for further morphological operations. The histogram is a graphical representation of the distribution of pixel intensity values in an image. A global color histogram is used to encode information from the image in fruit classification. It has a set of values, with each representing a distinct color. The statistical information is calculated using the global color histogram. The statistical information measures are mean, contrast, homogeneity, energy, variance, correlation and entropy. Sum and difference histogram method is used for image texture description. Border/Interior (BIC) classification is also used to describe images by separating the borders and interior regions (Patel et al. 2011).

Figure 26. Binary mask of an image is obtained using threshold segmentation for further morphological operations.

3.5.1.3. Regional Descriptors

Regional descriptors are used to compute the regional properties in an image. Regional descriptors like texture and moment invariants are calculated to extract regional properties in a binary image, as in Figure 27 (Riyadi, 2007). The region properties that can be calculated in an image are area, perimeter, centroid, primary axis length, and minor axis length, etc. The area of fruit in an image is measured using the total number of pixels in the fruit region. The perimeter of fruit in an image is calculated using the total number of pixels in the boundary region. Moment invariants are used to quantify the texture content in an image. Moment invariants used for textural analysis are mean, standard deviation, smoothness, third moment, uniformity and entropy. The fruit size can be easily identified using geometrical properties like area, perimeter, primary axis length and minor axis length. Other size parameters for fruit classification are a fruit's maximum and minimum diameter. These properties of fruit size can be easily measured using regional descriptors in image processing. The moment invariant helps to measure the texture of the fruit.

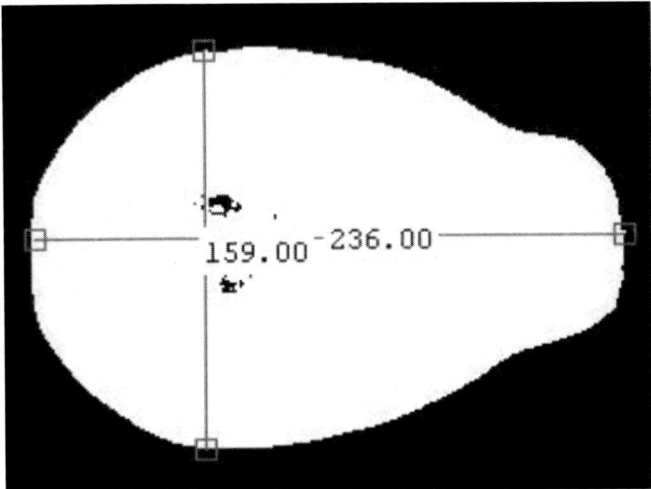

Figure 27. Regional properties calculated using the binary mask image.

3.5.1.4. Boundary Descriptors
The shape is an essential attribute in the classification of fruit. The shapes of fruit are generally irregular; most of the time, it is assumed to be its nearest regular shape. The characteristics of these fruits are determined by curve fitting. This curve fitting is time-consuming and the fruit standard is disturbed in this process. As an alternative, boundary descriptors can be used to determine the shape of the fruit. The boundary signatures can also be used to determine the shape of fruit.

3.5.2. Role of Features and Classifiers in Fruit Classification
Feature extraction, pattern classification and k-Nearest Neighbors algorithms were commonly used in fruit classification. Various features and classifiers are combined to classify fruit with higher accuracy. As features and classifiers are combined, the training data requires less training. Classifiers compare input data with trained data and provide classification results for the fruit.

Commonly used classifiers are support vector machines, neural networks, fuzzy and rough sets, Bayesian and evolutionary computing.

3.5.2.1. Pattern Classification and Nearest Neighbor Classifier

Pattern classification automatically identifies different objects in an image and extracts the required features from an image. Supervised classification is a set of feature vectors that characterize the boundary separating fruit. Unsupervised classification is a set of feature vectors for which no characterization or training is given in prior. K-Nearest Neighbors algorithm is also used to identify the input data by comparing it with trained data. It uses Euclidean distance measures to measure the distance between points in input data and trained data.

3.5.2.2. Neural Network Classifier

The artificial neural network classifier provides a higher degree of accuracy when compared with other classifier models. Feed-forward neural network model, multilayer perceptron is preferred in fruit classification. This model uses a supervised learning technique, back propagation neural network (BPNN), to train the data sets. The primary network structure of BPNN has an input layer, a hidden layer and an output layer. The network is trained with many input sets to predict the exact output result. The classifier uses different structures for each category of classification.

3.6. Tea Quality Assessment

Tea is one of the most regularly consumed beverages throughout the world. Tea production is done throughout the year from different tea varieties. Quality assessment of tea is determined through the color of tea leaves. Traditionally, tea process stages are determined by manual experts in the field using leaf color. Traditional assessment of tea quality and process stages is time consuming and based on hypothesis. The disadvantages of traditional methods have led to the use of recent innovations like image processing in the tea sector. The use of image processing techniques in quality assessment in different sectors has paved the way to implement tea quality assessment with easier, reliable and higher accuracy. In the tea sector, image processing can solve various problems and assist in decision making. Image processing methods are used to identify the endpoint of fermentation by monitoring tea leaf color, estimating the quality of tea at a drier mouth and minimizing manual inspection by introducing electronic means.

3.7. Sugar Cane Leaf Area Measurement

Sugarcane, the world's largest crop belonging to the grass family, is an economically important plant for the production of sugar. Sugar is demanded throughout the year in the entire world. This has enabled sugarcane industry and farmers to concentrate on increasing production with high quality and minimum cost. Leaf area assists in detecting disease and pest-infected symptoms, nutritional deficiencies, and lack of fertilizers and so on. The measurement of the leaf can be performed either through direct or indirect methods. In the direct method, all the leaves are measured individually; in the indirect method, destructive methods are used for measurements. These measurement methods are time-consuming, destructive and tedious job. The use of precision farming has made it possible to measure plant leaf area using the grid counting method. The acquired image of the sugarcane leaf is in the RGB color model. RGB color model image is converted into the grayscale image model. The grayscale image sets the threshold value to separate the leaf region from the background. The boundary of the leaf region or vein region can be determined by converting the input image into a binary image model (Patil & Bodhe, 2011).

3.8. Paddy Crop Growth Analysis

The monitoring of paddy crop growth from its infancy is very much important about its relationship with meteorological variability. Various traditional methods in monitoring crops and that of monitoring crops vary to individual fields. Traditional methods are inconsistent, unreliable and vary according to personal decisions. Recent innovations like remote sensing, GPS, and GIS are useful in monitoring paddy crops (Le et al. 2008). Satellite remote sensing uses optical sensors to survey the seasonal changes occurred in paddy (Kaizua & Imou, 2008). Optical sensors use a range of scales from pixel to region with a spatial resolution of low-to-medium. A portable spectral radiometer is also used in remote sensing to monitor the field with a higher frequency rate. The major disadvantage of the portable spectral radiometer is cost and time consumption. Another means of monitoring crop seasonal growth crop is by using digital came to collecting scientific data of quantitative information related to crop. Only a few digital camera models are available in the commercial market for image acquisition to assess crop growth. RGB-cam (color camera) and NIR-cam (Near infrared camera) are best suited for monitoring crop growth. The NIR camera is best suited for three-dimensional images of crops. This camera makes the geometrical measurements like plant length, and dry matter weight easier and faster. RGB camera was used to monitor the crops during day time and NIR camera was used to capture the images in nighttime (Lamb & Brown, 2001; Sakamoto et al. 2011).

Chaptaer 4

An Overview of Fundamental Digital Image Processing Procedures

The basic modules of image processing that determine the overall accuracy of image processing application in agriculture are discussed briefly in this section (Figure 28).

4.1. Acquisition

Image formation involves acquiring an image (through any capturing device) and storing it in a digital form (as an image matrix). An image, in general, is represented by two-dimensional function as f(x, y) where f is a function for the coordinate values of x and y. An image's function must always be nonzero and finite, $0 < f(x, y) < \infty$. The positive scalar value of a function is determined by illumination from the source and reflection or transmission from the object. The function of an image is in a continuous form and is converted to the digital (discrete) format for analysis. Image digitization is made possible through sampling and quantization (Butz et al. 2005).

4.2. Enhancement

Image enhancement is a manipulation process performed on an image to improve the quality and to have enhanced information from an image to satisfy the end user's requirements as in Figure 29. It effectively displays or records data for successive visual understanding. This process does not change inbuilt information about data but changes the dynamic range of specific features for localization (Costa et al. 2011). Most of the pre and post-processing algorithms are developed based on these image enhancement techniques. These methods are broadly classified into spatial and frequency domain techniques. Spatial enhancement techniques perform direct manipulations on the image pixels, whereas frequency enhancement technique performs indirect manipulation on images by making a transformation in the images either by using convolution or kernels.

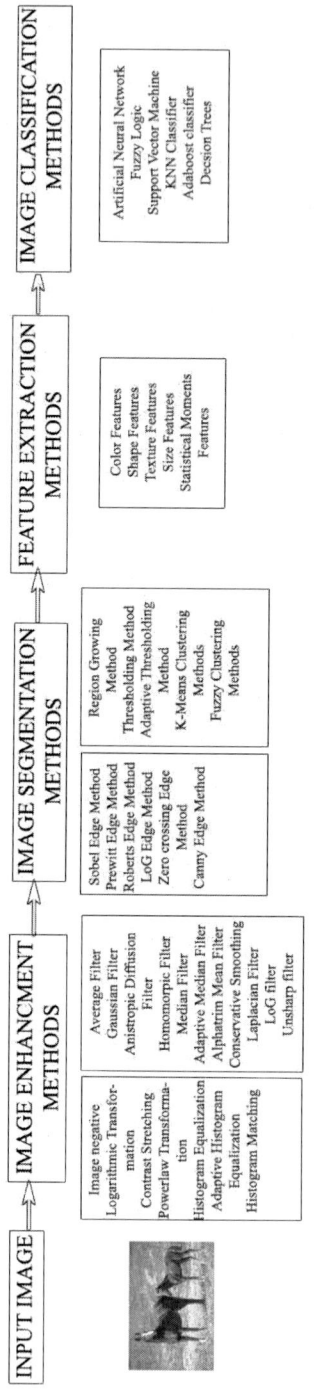

Figure 28. Basic modules of image processing.

An Overview of Fundamental Digital Image Processing Procedures 43

Figure 29. (a) Distorted input image and (b) enhanced output image.

4.3. Segmentation

Image segmentation is an essential module of middle-level processing used to segment regions of interest from the background (Figure 30) (Farmer & Jain, 2005). It is an essential step which helps to detect fruit and also assists in identifying the fruit. Numerous segmentation methods are involved in detecting regions like edge detection, thresholding, histogram, level set, Markov random fields and clustering methods (Camargo & Smith, 2009). Edge detection is a commonly used method for detecting the edges of an object. Edges are identified based on an image's abrupt change in pixel value. This basic concept of segmentation is calculated based on detectors developed using gradient values. Sensors such as Robert, sobel, prewitt, LoG, zero crossing, gabor, and canny are frequently used edge detectors. The histogram is used to measure the pixel distribution in an image. Global histogram, Local histogram, adaptive histogram, and histogram equalization are some commonly used methods for pixel distribution measurements. Adaptive histogram equalization is a common technique used to enhance the contrast in an image. This enriches the quality of the image and eases image analysis (Suryaprabha & Satheeshkumar, 2021).

Figure 30. (a) Input image and (b) region of interest segmented image.

Based on histogram distribution in an image, the threshold is set to identify the region of interest (ROI) from the background. Pixel values are similar for the constant color region in an image. This helps to differentiate ROI by having a constant pixel value from the background. Global thresholding, local thresholding and adaptive thresholding are standard methods used in object selection. Supervised and Unsupervised methods of clustering based on machine learning techniques can be applied to differentiate the foreground from the background. The most commonly applied clustering method is the conventional unsupervised k-means clustering method. A number of cluster selections is an essential feature in this algorithm. Markov random field is a Bayesian approach for labeling an image's region. A predefined set is fixed to similar label pixels in an image. Markov chain is a chain of random variables that depends on the preceding variable. Variable is the pixel value in an image that depends on the neighbor pixels. The same labels are marked for similar pixels based on pixel appearance with the nearest neighbor relationship.

4.4. Morphological Operations

The morphological operator is used for extracting objects in an image from which shape and size can be analyzed. The term morphology defines the physical structure of an object. Identifying physical structures helps mark the physical shape of objects in an image.

Identification of an object is only a part of the morphological operation. The basic operation of morphological operation is to represent the objects in an image and to describe them. Morphological opening and closing are used for smoothing and cleaning the boundary of the objects available in an image by opening and closing holes in the contours of an image as in Figure 31. Opening and closing are two contrast terms in morphological methods. Contours with small, narrow and weak edge-strength are broken in opening operation with the help of structuring element. Contours with narrow, thin and weak edge strength are combined in closing operation with the help of structuring element. The opening operation uses erosion followed by dilation.

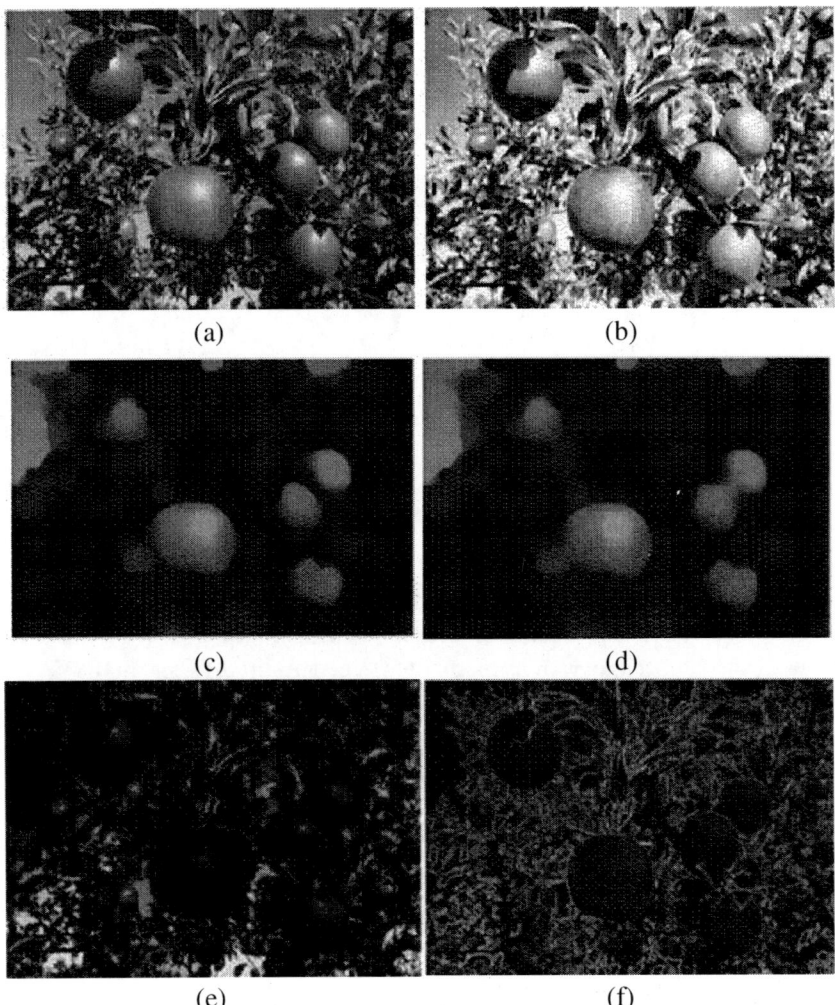

Figure 31. (a) Input image, (b) gray scale converted image, (c) opening operation applied output image, (d) closing operation applied output image, (e) and (f) morphological operation applied output image with different filter masks

The closing operation uses dilation followed by erosion. Erosion and dilation are the two essential concepts of morphological methods. A structuring element is a set of elements used for the manipulation of erosion and manipulation operation (Femat-Diaz et al. (2011).

Morphological filter masks are used in performing dilation and erosion operations. Kernel with 3 x 3 window is used as filter mask to perform the morphological operation. Three properties can be derived from the eroded and dilated image by performing image subtraction. The third property output obtained by subtracting dilated image from eroded image had a better object identification when the morphological operation was applied to an input image (Gao et al. 2009). The output of morphological operation and properties vary based on the filter mask selection. Boundary extraction, hole filling, extraction of connected components, Convex Hull, Thinning, Thickening, Skeletons and Pruning are the widely used morphological algorithms for object identification and description.

4.4.1. Image Representation and Description

Shape description is a process of extracting a required feature from an image. Objects in an image can be represented in a numeric form. The shape of the fruit in an image is transformed into a graphical representation. Statistical moments, boundary descriptors and regional descriptors are standard methods used for describing the image properties. The other statistical moments like mean, variance and smoothness be used to measure the color features in the objects in an image. Area, perimeter, eccentricity, Euler numbers and Euclidean measures extract size features in an image. Chain code, polygon, invariant moments, B-spline, fourier descriptors and shape descriptors are used to extract an image's shape features. Chain code is one of the simple methods to classify objects in an image. It is a directional code that uses fragments of straight line based on the direction and length of the line. Segments of linear line code used to form boundary in a unique direction are referred to as a freeman chain code. Labels are assigned in all the directions of boundary lines. Starting point determines the accuracy of the chain code. Fourier descriptor uses the Discrete Fourier Transforms and extends the functionality to identify the shape of objects in an image. Multiple descriptors are combined to determine the shape of an object. Planar shapes, closed boundaries and open curve shapes can be easily extracted from Fourier descriptors. Skeleton is a critical approach that reduces the object into a thin graphical representation. Medial axis transformation defines the skeleton of an object in an image (Moreda et al. 2012).

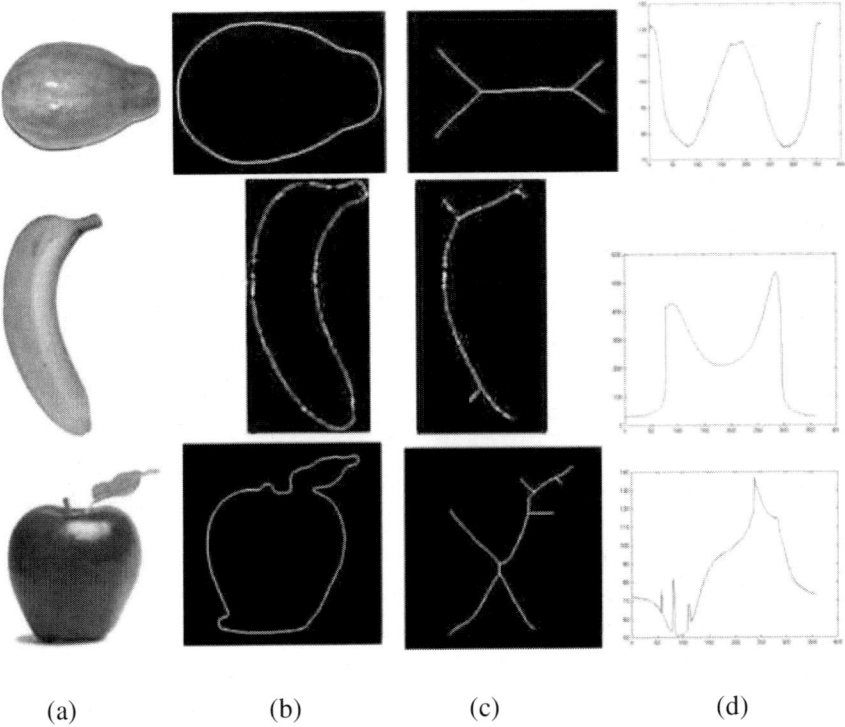

(a)   (b)   (c)   (d)

Figure 32. (a) Original Image, (b) Fourier descriptor applied output image, (c) Skeleton applied output image, (d) Shape signature applied output image.

Object region is identified by calculating the distance between an image's interior points and boundary points.

Shape signature is used to represent the shape of an object in the form of a 1D function. Shape signatures are formed by plotting points from the centroid to boundary distance. Another method for shape signature is plotting points from Eigen axis far away from the centroid. Eigen axis points are determined using a chain code algorithm. Centroid to boundary distance plotting is the simplest and best method for recognizing the shape of an object, as in Figure 32. This method can easily identify straight, slightly curved and curved boundaries. Rotating and scaling transformation methods can be applied in these boundary signatures, but this method is invariant to translation. Shape signatures and skeletons-based approach is always suggestible for better performance sorting objects in an image.

4.4.2. Object Recognition

Acquired, preprocessed, segmented, identified and recognized object enters object recognition as in Figure 33, the higher level concept of computer vision. Object recognition is used to recognize objects based on some learning and training concepts. Patterns and classifiers are the tools applied in object recognition. Patterns use specific descriptors to describe the objects recognized and classified based on these pattern recognizers. Vectors, Strings and trees are a common collection of pattern recognizers in this method. Pattern matching is an object recognition technique used to match objects based on their classes. Classifiers are used for the purpose of matching objects. These classifiers are also broadly classified into supervised and unsupervised methods. Numerous classifiers are available for pattern matching, like minimum distance classifiers, statistical classifiers, Bayesian classifiers and neural network classifiers. A minimum distance classifier is a simple method as it uses the concept of Euclidean measure to calculate the distance between the unknown and predefined patterns (Shih, 2010).

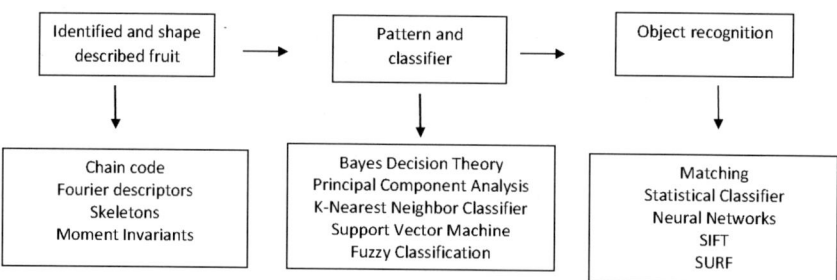

Figure 33. Concepts and methods of object recognition.

Bayes decision theory is an important concept that supports the complicated decision-making task. Decision-making situations are uncertain and unpredictable. It may be required to make simultaneous decisions within a short time. Bayes decision theory is a statistical concept of making decisions using probability and cost effect on the decision. In fruit sorting, Bayes's decision theory is used to sort fruits like lemon, amla, apple and banana.

Principal component analysis (PCA) is a commonly preferred statistical method for recognizing objects in a higher-level concept of computer vision. PCA is a pattern recognition method used in face recognition, image compression, and various pattern recognition applications. In fruit sorting, PCA is used to classify and categorize the fruits in a fully automated computer vision system. It uses statistical tools like standard deviation, Eigen vectors, and covariance matrix. Support Vector Machines (SVM) is used to label the objects in an image using a supervised and unsupervised learning technique. SVM can be trained to recognize objects in real time environment. SVM is a popular method for biomedical applications and can also be integrated to work with other applications. A neural network is a machine learning technique to classify different objects in an image. This method is inspired by the neural structure of the biological nervous system. Scale Invariant Feature Transform (SIFT) is an object recognition method that is trained to compare the objects with predefined and stored objects in a database. Key features are extracted from an object and are stored in the database. Each object present in the reference image is compared with the predefined objects in the database. Euclidean distance measures of the vector are applied to match the objects in the image. Speeded Up Robust Features (SURF) is an object recognition method used to detect and describe the objects present in an image. This method is faster, more accurate and highly efficient than the SIFT method in object recognition.

Chapter 5

Image Acquisition Methods Appropriate for Various Agricultural Situations

Image acquisition is an initial phase where the required object of interest is captured through different image acquiring devices like digital cameras, sensors, hyper spectral imaging, and infrared imaging and is stored in a system for further analysis and information interpretation. The process of obtaining a digital representation of a scene is called image acquisition. The output of an image acquisition process is the image that is composed of pixels. The digital tool or device used to obtain a scene is called a camera or imaging sensor. There are two commonly used technologies, such as i) charge-coupled device (CCD) and complementary metal oxide semiconductor (CMOS) are there for wide application in many sectors (Russ & Woods, 1995). The process of image acquisition involves i) capturing of light wavelength by analogic sensors as the major or minor charge that depends on incident light emitted from scenes, ii) amplification, filtering, transportation and enhancement of emitted light wave signals through particular hardware. A lens and an output interface are significant elements that perform the above tasks. Each image acquisition technology has some advantages and disadvantages that can be used for specific purposes. CCD equipment has features like high-quality images with less noise, operates under less or high illumination conditions and has better color depth. The special features of CMOS technology are processing images at a faster rate, needing less power supply to operate, having the option to choose the region of interest and its cost being cheap. The image acquisition for CCD and CMOS systems are recently implemented through Time delay and Integration (TDI) mode. The CCD or CMOS in TDI mode acquires images at high speed and operates at low light illumination conditions.

Image acquisition in TDI mode is commonly used for agricultural monitoring, field inspections, grading and remote sensing. The seven different imaging systems, such as 1. Mono-RGB vision systems, 2. Stereo vision systems, 3. Multispectral and hyperspectral camera, 4. Time of Flight cameras (ToF cameras), 5. LIDAR, 6. Thermography and fluorescence imaging, and 7. Tomography imaging can be helpful for agricultural purposes. Among them very commonly used for crop monitoring is an RGB camera during day time and a near infra- red (NIR) ToF camera during night time (Li et al. 2014; Fahlgren et al. 2015).

5.1. Mono-RGB Vision Systems

It comprises the following parts: input interface, lens, imaging sensor, hardware system and output interface. There are two mono RGN vision systems: line camera or matrix camera. It is very commonly used in many computer vision systems for morphological analysis. The successful application of this vision system was reported for yield prediction of black pepper, various phenotyping studies, and QTL mapping studies of maize. This system has advantages like generating more images in a short period at a lesser cost, being highly portable and suitable for many image processing algorithms coded on C++, Python, Java, Python and R. The disadvantages of this system include being sensitive to illumination, especially in outdoor situations. Light detection and ranging (LIDAR) devices are commonly combined with Mono-RGB vision systems to neutralize their disadvantages.

5.2. Stereo Vision System

It is the improved version of mono-RGB system and mimics human vision through incorporation of two mono vision system. Images acquired as known as depth maps which enables distance measurements. This system is useful for analyzing plant canopy features, to get three dimensional model plant can be well documented through this system. The successful implementation of stereo vison system that is two identical RGB cameras demonstrated to perform better in canopy analysis on grapevine, cotton and rapeseed (Chaerle & Van Der Straeten, 2001). The advantages of stereo vison system are i) simple architect, ii) enables to obtain depth maps, iii) can be used to obtain multi-view stereo (MSV), iv) MSV at a lower cost than LIDAR imaging. However, it is more sensitive to illumination, especially in the outdoor environment, and needs high performance algorithms, especially for image segmentation.

5.3. Multispectral and Hyperspectral Camera

These cameras are widely used in several scientific and industrial sectors. The spectral resolution is the key component that distinguishes the multispectral and hyperspectral images. Multispectral cameras acquire images at 3 to 25 spectral bands. The bands are not continuous in the case of multispectral images. However, hyperspectral cameras acquire pixels at nanometer wavelength levels. Additionally, hyperspectral images contain many continuous bands. These cameras are helpful in remote sensing field. Also, the multispectral images are valid for fruit recognition (IMEC, 2008). Hyperspectral cameras are helpful for pest, disease, or physiological defect, soil character determination, photosynthesis efficiency, leaf water content, Nitrogen status, leaf chlorophyll content and soil degradation by erosion (Blasco et al. 2007; da Silva, 2016).

5.4. ToF Camera

This camera is helpful for automatic plant recognition and phenotyping. TOF camera work on the principle that differences in the time taken to receive a signal emitted as near infrared (NIR) by an object (Tucker, 1979). This camera is more precise in 3D reconstruction. The ToF-mounted robots are demonstrated to identify QTL phenotyping on sorghum crops. The combination of stereo vision with TOF can solve image segmentation issues in leaf areas. Microsoft Kinetic is a low-price 3D imaging system popularly used for gaming purposes. The kinetic camera with perfect segmentation algorithm shoot or root architecture analysis (Lobet et al. 2011). However, ToF cameras have disadvantages like low resolution, the working distance (limited to a few meters only) and more dependency on reflecting surfaces. It has more potential in indoor plants and crops grown under protected structures.

5.5. LIDAR

It is an imaging technique developed for the remote sensing field. It uses laser pulse light and measures the time between emission and detection of reflected light for calculating the distance between light source and object. It forms a

3D structure of the objects available at distances from a few centimeters to thousands of kilometers. LIDAR is fixed in satellites for measuring the vegetation area, height, volume and biomass. Nowadays, LIDAR is now fixed in drones, manned and unmanned flights to get satellite images. It is well demonstrated to detect weeds in maize fields and phenotyping cotton and tomato. Complete monitoring of crop plants is possible through LIDAR. It has many drawbacks such as the requirement of calibrations, absence of color measurement, taking more time to compute, being less accurate during massive phenotyping, causes scanning noises due to heavy wind, rainfall and movement of insect-like organisms.

5.6. Thermography and Fluorescence Imaging

The thermography imaging technique is very commonly used in remote sensing and plant monitoring. This camera acquires images at 300 – 14000 nm wavelengths, which lets measurement of temperature through conversion of irradiated energy. It is commonly used for irrigation management in agriculture by measuring evapotranspiration at the level of a single plant to the field or whole region. Thermography is also useful for forecasting disease development in relation to temperature. Thermography is sensitive to temperature, wind speed, irradiation, and leaf or canopy position that affect the quality of the images. The UV light (340 to 360 nm wavelength) reflected by plant parts is used in fluorescence imaging. These techniques are useful in agriculture to measure plant components such as amino acids, chlorophyll contents, biological processes like photosynthesis and secondary metabolites (Sirault et al. 2009). The combination of thermographic and fluorescent imaging needs neural network-based computation techniques, which can solve many emerging problems (Jones et al. 2002).

5.7. Tomography Imaging

Tomography imaging techniques such as Magnetic resonance imaging (MRI), Positron emission tomography (PET), X-ray computed tomography (x-ray CT) and High-resolution x-ray computed tomography (HRXCT) are non-destructive imaging techniques widely used in the Medical sector but also having potential for agricultural purposes. MRI uses a radio frequency magnetic field to acquire tomographic images. In agriculture, MRI imaging is useful for studying plant metabolites, physiological processes and the maturity process of seeds or grains. PET is also helpful for understanding biochemical processes, monitoring and diagnosis diseases. In plants like sorghum, it is used to study the role of carbon in photosynthesis. X-ray CT technique generates 3D images of the specific internal part of plants. In addition, this can be used to study root systems or architecture

Chapter 6

Image Enhancement Tools for Images Acquired in Different Agro Conditions

Image enhancement is the foremost primary step in image processing applications. It is used to improve the image quality to make it understandable for machine and human perception. The quality of an image determines the accuracy of information retrieval and interpretation from an image. Improving the image quality to its highest standard is essential for better perception. Generally, image quality enhancement can be achieved either by reducing the noise or adjusting the contrast, or modifying the brightness of an image. It does not change the inherent nature of the image. Still, it limits only the dynamic range of pixels which ascertains the smallest and highest intensity value in an image. Two main categories of image enhancement techniques are spatial and frequency methods. In the spatial domain, images are directly manipulated using their pixel values; alternatively, in the frequency domain, images are indirectly manipulated using transformations (Gonzalez, 2009; Charbit, 2010).

The spatial domain technique performs direct manipulation of the image pixels. The term spatial is a representation of an image plane; comprising intensity values for every pixel in an image. This method operates directly on each pixel value and these values are enhanced by mathematical manipulations on individual pixels. The manipulation can be done based on point processing or neighborhood processing. The gray level transformation or point processing, has a 1x1 neighborhood to manipulate the image using,

$$s = T(r);$$

where s is the enhanced image for the input image r and T is the manipulation performed on the input image.

Some fundamental gray level transformations are image negatives (compliment), log transformation, power-law transform, piece-wise linear transform, gray level slicing and bit plane slicing. If each pixel value in an image is enhanced with the support of its neighboring pixel values, then the methods are termed neighborhood processing methods. It is also referred to as spatial filtering as it has a larger neighborhood to perform a predefined operation to enhance the image. This predefined operation is manipulated using spatial mask, filters, kernels, templates and windows. The nature of the filter is determined using a spatial mask with a predefined operation. The resultant filtered image is attained when the center point of the filter mask visits each pixel in the input image. The operation performed on the image pixel is linear; then, the filter is referred to as a linear spatial filter; otherwise, it is referred to as nonlinear spatial filter (Geman et al. 1990).

Correlation and convolution are the two important terms in spatial filtering. Correlation determines the degree of likeness among data sets. Correlation is a process in which the kernel is moved over entire pixels in an image and the predefined manipulation is performed in each pixel position. Image convolution is similar to correlation, except kernel is rotated to $180°$ before performing manipulations. The image convolution operation is used widely for enhancing effects (like blurring, edge sharpening etc.) in the images.

The frequency domain technique performs indirect manipulation of images by transforming the images using convolutions or window kernels. In some instances, the image can be processed better in the frequency domain than the spatial domain. The image is first converted to the transform domain in the frequency domain. The manipulations are performed on the transform domain using forward transformation. After manipulation, it is again converted to the spatial domain by doing the inverse transformation. The resultant image in the spatial domain will be better and improved to satisfy the end user's need. Thus, in frequency domain, image position modification results in spatial domain changes. Alterations in the spatial domain are based on the rate of change in intensity values in an image. Various transforms available in the frequency domain include Discrete Fourier transforms, Fast Fourier transforms, Discrete Cosine transforms, Continuous wavelet transforms, Windowed Fourier transforms, Walsh Hadamard transforms, Haar transforms, Hotelling transforms and so on.

Discrete Fourier transform is one of the vital transform models used in image processing. It represents the image in the frequency form. It plays an essential role in the process of image filtering. It is an image processing technique used to smooth, de-blur and restore the image to an enhanced form. The filtering process is performed either as a low pass filter or a high pass filter. A low pass filter reduces the amplitude of high frequencies and has low frequencies unchanged. This filter is very much helpful in smoothing the images. A high pass filter retains high frequencies and soothes the amplitude of low frequencies. This filter is very much helpful in sharpening the images. Theoretically, the spatial domain could be seen easier for smoothing and sharpening operations on the images. But in practice, the frequency domain is commonly used for grinding and sharpening operations on images [70]. The filter mask is an operation carried out on the image array set. The image array set has a set of scalar components multiplied by the Fourier term's amplitude. The role of filtering mask in transformation can be represented mathematically as,

$$G(m,n) = H(m,n) * F(m,n)$$

where $F(m, n)$ is the input image, $H(m, n)$ is the filter mask and $G(m, n)$ is a resultant enhanced image. The role of the filter mask is significant in noise removal and edge detection. The filter mask helps to make the image understandable, clear and sharper. Different mask models like LoG operator, Roberts operator, Sobel operator, prewitt operator and so on are used for noise removal and edge detection technique (Solomon & Breckon, 2011).

6.1. Point Processing Methods

It is a simpler and easier technique to enhance image quality by manipulating pixel-by-pixel an image. In point operation, a mathematical equation or operation is applied over every pixel point expressed as,

$$A(s,t) = T[B(s,t)]$$

where B(s, t) is an input image and T is a point operation applied over each point in an input image to result in an enhanced image $A(s, t)$.

Again, this technique is broadly classified into two categories: gray level transformations and histogram processing as in Figure 34. Mapping each pixel value in an input image to a pixel value in an output image using a transformation function is termed intensity or gray level transformation. Mapping the number of pixel occurrences for a particular intensity value in an image is called histogram. The manipulation performed on a histogram using a discrete function is termed Histogram Processing (Starck et al. 2003).

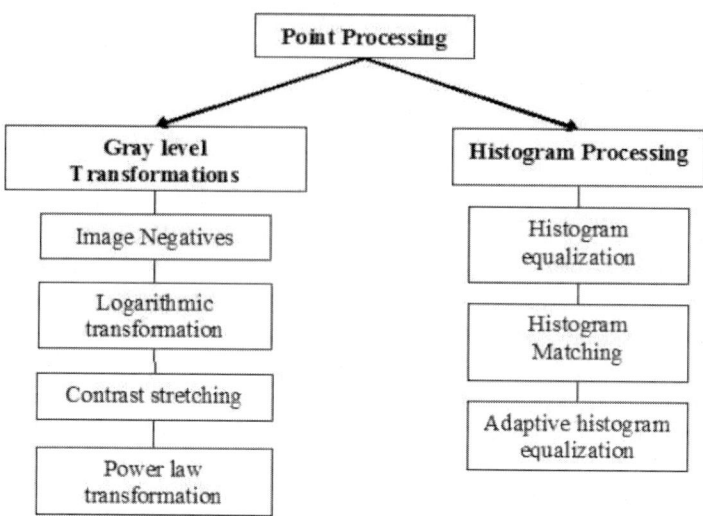

Figure 34. Different types of point processing methods.

6.1.1. Gray Level Transformation

Gray level transformation, also known as intensity transformation, is a function used to map a pixel value in an input image to a new pixel value in the output image using the transformation function. These gray-level transformations are either linear or non-linear. In linear gray level transformation, it uses a linear function for mapping pixel values at same location in the images. The non-linear gray level transformation uses a non-linear function to map the pixel values. Image negatives and contrast stretching are some examples of linear gray level transformations. Logarithmic transformation and power law transformations are some examples of non-linear gray level transformations. These intensity transformations are not performed directly in RGB image color space and during manipulation, the input image in RGB color space is converted into gray scale color space.

Logarithmic Transformation is commonly used to compress or expand the dynamic series of pixels in an image's light or dark regions. Hence, it is required to map lower intensity values with a higher range of grayscale values and higher intensity values with the lower range of grayscale values, as in Figure 35. Pixel values in an image are transformed or replaced using the logarithmic value of each pixel using the formula,

$$x = c * \log(1 + a)$$

where c is a scaling constant used for image quantization and a is an intensity value of the original image (I) at a point (m, n).

The increase in constant value increases the image brightness. Therefore, choosing appropriate constant value for enhancement is essential to avoid ill effects like blurriness. The numeric value one is added in the scaling factor calculation to avoid problems in situations where log value is undefined.

(a)   (b)

Figure 35. (a) Contrast distorted input image and (b) Logarithmic transformation applied output image

Power Law Transformation is an alternative to logarithmic transformation. It raises the pixel values in an image to a fixed power and is calculated using,

$$K(s,t) = c\bigl(I(s,t)\bigr)^{\gamma}$$

where $K(s, t)$ is the enhanced image, $I(s, t)$ is an input image with c and (γ) are positive constants. The constant c is used for scaling, and λ is the exponent used to improve image contrast.

If the gamma value (γ) has a value larger than 1 then it improves the image contrast of the light region in an image otherwise, it enhances the contrast of the dark region in an image, as in Figure 36(a). This method is useful to manipulate the contrast of an image for general purposes. Several images capturing devices, printers, scanners and monitors use the concept of the power law for gamma correction, which is a correction required on the output of any of this device for higher display accuracy. Gamma correction is very useful for solving the non-linear relationship between voltage, considered as input, and intensity, considered as output, in a monitor display.

Piecewise Linear Contrast Stretching is a commonly used piecewise linear function method where the intensity level range of an image is expanded for normalization and thereby contains a dynamic range of value in the resultant output image, as in Figure 36(b). The input requirement is more in this method when compared with other gray level transformation methods where contrast stretching is given by,

$$K(s,t) = (I(s,t) - C)\left(\frac{(x-y)}{(u-v)}\right) + s$$

where $K(s, t)$ is the output image, $I(s, t)$ is the input image with four input values s, t, u and v. Values of x and y represent the upper and lower limits of the pixel range that are in use for quantization, i.e., $x = 255$ and $y = 0$ for an 8-bit image. Values of u and v represent maximum and minimum pixels that are present in the image. The value of u and v is ascertained from the input image histogram.

6.1.2. Histogram Processing

The histogram is used to plot the number of frequently occurring pixel intensities in an image. The horizontal axis of the histogram represents information related to the intensity values in a range of [0 − 255] for an 8-bit image. The vertical axis represents information related to the frequent occurrences of each intensity value within an image. Enormous information is gathered through histogram as it provides global information about the image's properties, appearance, and texture. In histogram processing, an image's contrast is enhanced by mathematical manipulations over the histogram. It modifies an image's active range of pixel intensities using a discrete transformation function. It is used in different modules of image processing like image enhancement, image segmentation, image description and image compression and is mainly helpful in processing the real-time images. Histogram equalization, histogram matching and adaptive histogram equalization are some commonly available histogram processing methods.

(a)　　　　　　　　　　　　　(b)

Figure 36. (a) Power-law transformation applied output image and (b) Contrast stretching applied output image.

Histogram Equalization is a standard method used to enhance the image contrast by spreading the intensity values evenly in an images in Figure 37(a). The complete automatic process with the simple computational task is the major advantage of this method. The result of the enhanced output image is purely dependent on the input image's histogram. It considers the variable assigned to intensity values of input image either as a continuous or discrete variable. In this method, the acquired input image is remapped or transformed into a new image, i.e., the output image using the mapping function,

$$I = f(A)$$

where A is the intensity values of the input image and I is the intensity values of the output image. $f(A)$ must be a single value and increase monotonically, which are the major conditions determining the validity of mapping. It uniformly distributes the input image's histogram to obtain an enhanced output image. This mapping is done using the probability density function, which assures that the histogram of the output image is equally distributed.

Adaptive Histogram Equalization is global and is suitable for situations where the entire image region needs to be enhanced. It lacks to perform better for local region enhancement in an image. The histogram of local regions must be manipulated to enhance the required local region in an image, as in Figure 37(b). This process is achieved by adaptive histogram equalization, where the different regions in the image are manipulated through regions' local properties. The sliding window approach is a simple, easier and standard method used to enhance the image using adaptive histogram equalization. It breaks the image into different small blocks or tiles or windows and these blocks use the outer window to obtain the required histogram equalization. This method is very successful and helpful in increasing the contrast between local regions. There is a higher probability for over increase in contrast and occurrence of block artifacts in an image. To avoid these artifacts, the outer window size is increased comparatively to the inner window size. To restrict the increase in contrast value within a certain limit, the Contrast limited adaptive histogram equalization method is expanded from the adaptive histogram equalization method for better and more efficient enhanced results (Gilboa et al. 2002).

(a)   (b)

Figure 37. (a) Histogram equalization applied output image and (b) Adaptive histogram equalization applied output image.

Histogram Matching or Histogram Specification is based on the same principles of histogram equalization. Unlike histogram equalization, where the target histogram distribution is automatic, in this method, the target histogram distribution is user-specific. This technique is suited when the user has knowledge or idea about the regions in the input image that require enhancement. The shape of histogram needed is specified manually either by a mathematical function or from an existing reference image with the necessary histogram distribution as in Figure 38(a).

6.1.3. Fuzzy-Based Enhancement Using Fuzzy If-Then Rules

The developments and innovations in the concept of fuzzy logic paved the way for applications in image processing. This concept of fuzzy logic was initially integrated into image processing by researchers like Prewitt, Pal, et al. and Rosenfeld. The pixel values, the key constituent in an image, are uncertain, imprecise and indeterministic. So during the development of an automated system for real-time environments, the interpretations based on the crisp set of pixel values might mislead. So the use of fuzzy logic by considering pixel values as fuzzy would produce an accurate, specific and reliable result. Fuzzy image processing is considered a compilation of varied fuzzy approaches with three main stages: fuzzification, modification of membership values and defuzzification. The fuzzy set concept is applied in numerous modules of image processing like image enhancement, image segmentation and retrieval. In image enhancement, the contrast of an image is adjusted by modifying the membership values to transform the original image into the enhanced image, as in Figure 38(b).

(a) (b)

Figure 38. (a) Histogram matching applied output image and (b) Fuzzy based if-then method applied output image.

Fuzzy rule-based methods are instrumental even for problems that are non-linear. It is tedious to define deterministic criteria for enhancing an image. This task has been made simpler using the fuzzy approach. It is based on the simple classical rule system if (specific condition) then (specific action). Specific rules are defined for the pixels in an image for enhancement. These rules or conditions are formed by considering an image's gray level pixel value. Based on these conditions, decisions are made individually and then it is combined to make a final decision. In a simple fuzzy if-then system, an image's maximum, minimum and mid gray levels are calculated. As a fuzzification process, the membership values are assigned for an image's different (dark, gray and bright) regions. Then a fuzzy inference is made to modify the membership functions in an image. As a consequence of the inference mechanism, the pixel values of different regions with dark, gray, and white is transformed into black, gray and white. Then using the inverse of fuzzification, the result of the inference system is defuzzified (Pal & King, 1981; Pal & Rosenfeld, 1988).

6.2. Neighborhood Processing Methods

In neighborhood processing or spatial filtering method, an image is enhanced by applying a mathematical function to each pixel in an image and its neighboring pixels. Correlation and convolution are two important terms in spatial filtering. Correlation determines the degree of likeness among the data set in which the kernel is moved over entire pixels in an image, and the predefined manipulation is performed in each pixel position. Image convolution is similar to correlation except that spatial mask is rotated to 180 degrees before performing manipulations. The image convolution operation is used widely for enhancing effects (like blurring, edge sharpening, etc. in the images. The main idea behind spatial filtering is to combine the mathematical function and convolution mask for enhancement. This combined performance of convolution and mathematical function is technically termed filtering. It is mainly used for the purpose of eliminating unwanted or repetitive information in an image. Filtering techniques used to manipulate images in spatial domain are referred to as spatial filters, which are either linear or non-linear.

Linear filters modify the targeted pixel value by using specific linear combinations of the neighborhood pixel values, whereas non-linear filters use any non-linear combinations of neighborhood pixels for modification. The filtering mechanism is usually done by calculating the product of all pixel elements in the kernel along with its corresponding pixel elements in the neighborhood of the target pixel. Then all these products were summed up. The kernel, window, or mask (usually rectangular in shape) is moved over entire pixels in an image to perform the predefined operation in each pixel value to attain a newly enhanced pixel value for the whole of the image. Two main aspects considered essential in image enhancement are image sharpening and image smoothing as in Figure 39. In Image sharpening, image information is detailed so that it contains high spatial components of the image. In Image smoothing, image information is reduced because it has low spatial components of the image.

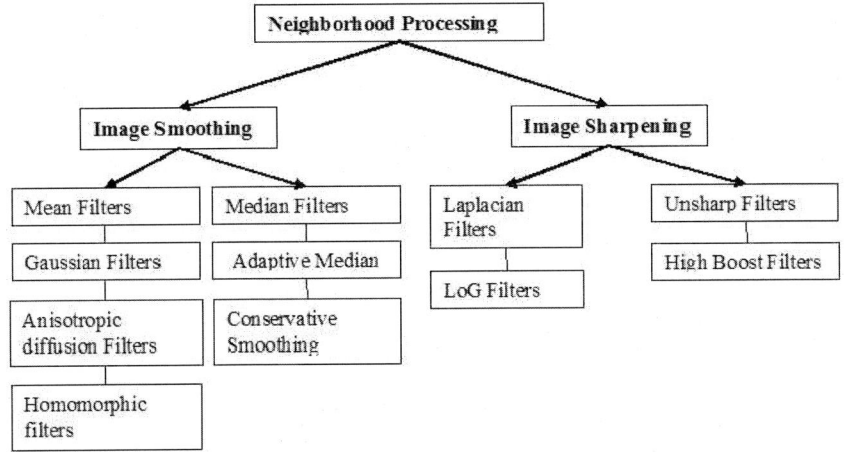

Figure 39. Different neighborhood processing methods.

6.2.1. Image Smoothing

Image smoothing is a local pre-processing method used to suppress the noise (i.e., unwanted information) and to blur (i.e., remove small details) the image. It is very much effective to remove impulse noise in an image. It reduces the high-frequency components in an image and retains low-frequency components to smooth the image. It removes the higher frequency information by blurring the image. Image smoothing has more effect on smoothness as the window size of the mask is increased but has a disadvantage that the required image features are removed. The input image in RGB color space must be converted into a gray-scale to make the image suitable for pre-processing. In this chapter, the low-level distortion of the blur effect is focused and these images are considered for analysis. There are different linear and non-linear filters used to enhance the image using smoothing filters and some of these commonly used filters are as follows;

Mean or Average Filtering is the simplest linear spatial filter used to smooth images for noise suppression by reducing the variations in pixel intensities, as in Figure 40(a). This method is similar to the convolution concept used in the spatial domain. The filtering kernel is square, like 3x3, 5x5 for sliding over the entire image. In this technique, the center pixel value in the kernel is replaced by a new pixel value as the mean of its neighboring pixels. This method effectively removes noise, but high frequency information in the image is lost. Image quality is degraded when large kernel is applied over the image (Greenberg et al. 2002).

Gaussian Filtering is a linear spatial filter used for smoothing, thereby reducing the noises in an image like mean filters as in Figure 40(b). This method is best suited for Gaussian noise removal and uses a bell-shaped 2-D Gaussian kernel to change the pixel values. The 2-D Gaussian function is defined as,

Figure 40. (a) Mean Filtering applied output image and (b) Gaussian filtering applied output image.

$$f(s,t) = \frac{1}{2\pi\sigma^2} e^{-(\frac{s^2+t^2}{2\sigma^2})}$$

where σ is the standard deviation for a function f with coordinates, s and t.

Standard deviation and kernel size are two major parameters in this filter. Standard deviation controls the smoothing result where larger the standard deviation value requires a larger convolution kernel. This value suppresses high-frequency information in an image and is referred to as low pass filters. In contrast to the mean filter, it uses a weighted average in the kernel to change the center pixel value. This Gaussian filter is considered as a primary step in a canny edge detector for noise suppression and is well suited for analysis in the frequency domain analysis (Ko & Lee, 1991).

Anisotropic Diffusion Filtering or Persona-Malik diffusion is a nonlinear and space variant partial differential equation (PDE) proposed by Persona & Malik to enhance and segment image. Preservation of edge without any information degradation in an image is the main advantage of this method, as in Figure 41(a). The PDE-based anisotropic diffusion equation is expressed as,

$$\partial k_i = div(A(||\nabla x||)\nabla x)$$

where x is a 2–D gray scale image. *Div* represents the divergence operator and represents smoothed image x at step i. $||\nabla k||$ is the image gradient magnitude and ($||\nabla k||$) is a function for function. If the function approaches zero, then the gradient magnitude value reaches its infinity and will not affect the edges. Hence diffusion process is done only in the interior region of an image to get a better-smoothed image. disadvantage of this method is that there are more probabilities for divergence during the evolving diffusion process. Bias has been added to the existing equation to resolve this issue. Another drawback is its inability to distinguish high gradient fluctuations in the noise from the edges (Gerig et al. 1992).

Rank Filtering. Order statistics or rank filtering is a non-linear spatial filter in which the center pixel value of the kernel is changed based on the ordering or ranking of pixels in the kernel region. These values are arranged in ascending order for performing neighborhood operations. Max, min and median are some common non-linear ordering filters considered for enhancement. Max filters use the 100th percentile value or maximum value of the kernel in the center pixel value of the kernel, whereas median filters use the 50th percentile pixel value and min filters use the 0th percentile or minimum value to change the center pixel value of the kernel.

Median Filtering. Median filter is a rank filtering method capable of reducing the irrelevant and vague information in an image. It also preserves high-frequency details in an image, unlike mean filters as in Figure 41(b). This method also uses square-shaped kernels like 3x3 and 5x5 for changing the pixel values. The median value is calculated for the kernel and its corresponding center pixel value is substituted by this median value. The median value is calculated by sorting the entire pixels in the kernel either in ascending or descending order. Therefore this filter is also termed as order static filter or rank filter. This kernel is moved over the entire image to get an enhanced image for interpretation. Removal noise with less degradation in an image is the main advantage of median filtering. Computational time is expensive in this method as it requires the ordering of numerical data in the kernels. This method is very suitable for eliminating the impulse or salt and pepper type noise from an image.

(a) (b)

Figure 41. (a) Anisotropic filtering applied output image and (b) Median filtering applied output image

Adaptive Median Filtering is also a rank filtering method used to remove the impulse noise without reducing or removing any high-frequency information in an image. This is an efficient and advanced technique compared to the standard median filters. Preserving the detailed information of an image even during the smoothing process is the main advantage of this method. Unlike other filtering techniques, this method can changing or modifying the pixel kernel size and threshold criteria during the manipulation process. The major purpose of this method is to reduce the distortion and to remove the impulse noises by smoothing the image without any loss of information, as in Figure 42(a). Smaller the window size, the lesser the loss of information during the smoothing process in an image and vice versa for large windows in which there is a high loss of information during the smoothing process (Polesel et al. 2000).

Alpha Trimmed Mean Filter It is considered a non-linear filtering technique combining both mean and median filters. It is similar to average filtering in the manner it calculates the average value of pixels with its neighborhood in the kernel. It combines the concept of order statistics to improve the filtering performance. It uses the median filter to arrange the neighboring pixels in ascending order and to set an alpha value for trimming the pixels. This alpha value removes extreme minimum and maximum values in the kernel to have an optimized output from filtering, as in Figure 42(b). The alpha trimmed mean filter is performed using the formula,

$$ATMF = \frac{1}{(n^2 - 2\alpha)} \sum_{i=\alpha+1}^{(n^2-\alpha)} M_i$$

where M_i is a square mask with a size of $n \times n$ and α is any value within the range of 0 to the median. If the alpha value is zero, the filtering technique acts like a mean filter. If the alpha value is calculated from the below formula,

$$\alpha = (n^2 - 1)/2$$

then the filtering technique acts as a median filter. Alpha trimmed mean filter is best suited for images distorted by the type of Gaussian and salt & pepper noise.

(a)   (b)

Figure 42. (a) Adaptive median filtering applied output image and (b) Alpha trim filtering applied output image.

Conservative Smoothing is a filtering technique that preserves high-frequency information but has less noise suppression capability, as in Figure 43(a). This method is effective towards noise spikes but weak towards additive noise. It is a simple and fast filtering technique that takes the range of pixels between maximum and minimum value in the kernel, excluding the target pixel taken for comparison. If the target pixel lies outside the range and if it is greater/lesser than the maximum/minimum, then it is replaced by the maximum/minimum value else, the pixel value remains unchanged. Similarly, this concept is done over the entire image region to get a better-enhanced image.

6.2.2. Image Sharpening Filters

Image sharpening focuses on changing intensity value which gives more importance to the human perception in such a way that it provides more information about the image. Unlike image smoothing where image details are degraded, image information is highlighted in this method to improve the image quality. Higher frequency components are retained and the lower frequency components are removed using sharpening operation. Image sharpening filters are based on the idea of derivatives and are classified as first order and second order derivative based filters. First-order derivative filters are based on gradient and are non-linear in nature and second-order derivative filters are based on Laplacian and are linear. These linear and non-linear filters of image sharpening include methods like Laplacian, Sobel and Prewitt. These methods are commonly used to identify edges by finding the discontinuity or abrupt change in pixel values in an image and are discussed below in detail:

Laplacian Filters is a second-order derivative spatial filter that highlights regions with rapid change in intensity value in an image. Laplacian for a 2–D function A (s,t), in general, is expressed as,

$$\nabla^2 A(s,t) = \frac{\partial^2 A(s,t)}{\partial s^2} + \frac{\partial^2 A(s,t)}{\partial t^2}$$

This equation is convolved through the entire image using a spatial mask. The equation used for calculating the Laplacian for image enhancement by performing a convolution task is given as,

$$I(s,t) = A(s,t) + c[\nabla^2 A(s,t)]$$

where $I(s, t)$ is an enhanced image, $A(s, t)$ is the original image and c is a constant with a value of either 1 or −1. If the mask has a positive value in its center coefficient, then the value is 1 else, it is −1.

Laplacian is very much useful for image sharpening and edge detection. This filter is capable of producing a fine edge based on the change in gradient value. Laplacian is very sensitive to noise, so it is not suitable in isolation for better enhancement. So it is combined with other techniques to have an efficient enhanced result, as in Figure 43(b).

(a)   (b)

Figure 43. (a) Conservative smoothing applied output image and (b) Contrast stretching applied output image.

Laplacian of Gaussian (LoG) Filters The major drawback of sensitivity to noise in Laplacian is solved by combining the Laplacian kernel and Gaussian kernel to produce a LoG filter. Computational time is less in this method when compared with the Laplacian method. Filtering is first done with the Gaussian function and then with the Laplacian function as in Figure 44(a). The gaussian function is used to smooth (blurs) the image, and then the Laplacian function produces double edges to have an efficient filtering method. Thus in LoG filtering, both techniques are combined as a single kernel to have an efficient filtering method and are calculated as,

$$\nabla^2 A(m,n) = \left[\frac{m^2 + n^2 - 2\sigma^2}{\sigma^4}\right] e^{-\frac{x^2+y^2}{2\sigma^2}}$$

where σ is standard deviation for an image function A with co ordinates s and t.

Unsharp Filters, also known as boost filters, is an alternative enhancement technique to the Laplacian method for image sharpening. In this method, smoothed image is subtracted from its original image to enhance the high-frequency components like edges in an image, as in Figure 44(b). Smoothed image for this process is achieved either through mean or Gaussian filtering. Unsharp filtering is calculated as,

$$A_e(i,j) = A_o(i,j) - A_s(i,j)$$

where $A_o(i, j)$ is the original image, $A_s(i, j)$ is the smoothed image, and $A_e(i, j)$ is the unsharp filtered image. The resultant image is added back again to the original image to have higher degree of sharpness. It is referred to as a Boost filter, as this method boosts up the sharpness of an image and can increase the high-frequency edge details in an image. This final step is calculated as,

$$A_{en}(s,t) = A_o(s,t) - k * A_e(s,t)$$

where $A_{en}(i, j)$ is the enhanced image and k is a scaling factor used to ensure that the resulting image falls within the proper range. If the scaling factor is greater than one then this filtering is referred as high boost filtering method.

Homomorphic Filtering is a generalized non-linear filtering technique developed with the basis of the illumination reflection model. It uses a nonlinear technique to convert the signal into additive signals and a linear technique to enhance these signals. This filtering method is used in both spatial and frequency domains of enhancement. It supports contrast improvement and brightness stabilization in an image. This filtering is very much more effective for multiplicative noise than other noise methods. This filtering technique considers an image as a combination of illumination and reflectance and is defined as follows,

$$A(s,t) = L(s,t) * R(s,t)$$

where $A(s, t)$ is an image, $L(s, t)$ is the illumination, and $R(s, t)$ is the reflectance of an image.

It is not possible to manipulate separately on the illumination and reflectance factor. The logarithmic component of image intensity is taken for manipulation over the discrete Fourier transform for enhancement. The high pass filtering technique is considered in Fourier to transform to suppress the low-frequency information and to highlight the high-frequency information in an image. After the Fourier transformation operation, the image is again converted to its original form by doing exponential transformation as in Figure 44(c).

Figure 44. (a) LoG filter applied output image, (b) Unsharp filtering applied output image and (c) Homomorphic filtering applied output image.

Chapter 7

Image Segmentation Methods Concerned with Plant and Agriculture Images

Image segmentation is a process of segregating foreground objects from background objects. The foreground object is the required region of interest in an image. So after extracting the region of interest from an image, its background is subtracted. There are numerous image segmentation algorithms available in the literature. Image segmentation methods are broadly categorized into edge, region, and thresholding, as shown in Figure 45. In edge-based methods, the required region of interest from an image is segmented based on the discontinuity property of pixels. In region-based methods the region of interest from an image is segmented based on the similarity properties among the pixels. In thresholding-based methods, images are segmented using similar intensity values. These classical segmentation methods require manual input of information or value to segment an image's required region of interest. It is a tedious job to segment an image autonomously and independently without any manual intervention. So to ease this job mathematical numerical method has been applied to image segmentation. Partial Differential Equation (PDE), a numerical method for solving complex problems, is used for this purpose (Deepa & Sivanandam, 2011).

Segmentation based on PDE is done by solving PDE equations. The commonly available methods in the literature are active contour models and geodesic active contour models. These deformable methods perform segmentation based on the global properties of an image by considering either edge or regional properties (Civicioglu, 2013).

In certain literature, image segmentation algorithms are broadly classified under two categories boundary-based and region-based segmentation. Boundary-based segmentation methods are based on pixels discontinuity properties and region-based segmentation methods are based on pixels similarity properties. These methods of image segmentation are mainly based on two properties. One is similarity property and the other is dissimilarity or discontinuity properties of relationship among pixels. For similarity -based methods, pixel values with similar intensity values are considered the basic criteria. For discontinuity property-based methods, pixel values with rapid and unexpected changes are considered the basic criteria. Some examples of similar property-based methods are region growing, region split and merge, thresholding, and so on. Some examples of dissimilarity property-based methods are edge detection, corner detection, line detection, and so on. Hybridizing the boundary- and region-based method is competent to produce a better segmentation result.

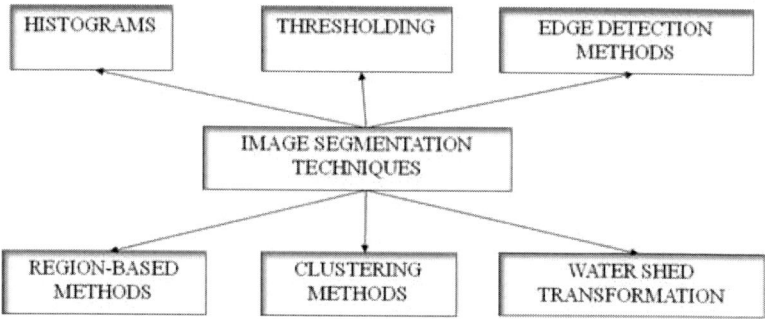

Figure 45. Different methods of image segmentation.

7.1. Edge-Based Segmentation Methods

Edge detection is a popular method of segmentation based on the property of dissimilarity values among the pixels in an image. It identifies the location of where pixel values change abruptly, and these locations are considered as boundaries of a region or an object in an image. These rapid changes in the intensity value of an image are considered edges.

7.1.1. Steps in Edge Detection

Edge detection contains three steps, namely Filtering, Enhancement and Detection. The overviews of steps in edge detection are as follows.

7.1.1.1. Filtering
Images are often corrupted by random variations in intensity values, called noise. Some common types of noise are salt and pepper noise, impulse noise

and Gaussian noise. Salt and pepper noise contains random occurrences of both black and white intensity values. However, there is a trade-off between edge strength and noise reduction. Filtering is performed to reduce noise, which results in a loss of edge strength.

7.1.1.2. Enhancement
To facilitate the detection of edges, it is essential to determine changes in intensity in the neighborhood of a point. Enhancement emphasizes pixels with a significant difference in local intensity values and is usually performed by computing the gradient magnitude.

7.1.1.3. Detection
Many points in an image have a nonzero value for gradient, and not all of these points are edges for a particular application. Therefore, some methods should be used to determine which points are edge points. Frequently, thresholding provides a criterion used for detection.

7.1.2. First Order Derivative Based Methods

In first-order derivatives, these functions are defined based on the criteria that they must be zero in areas of constant intensity. They must be non-zero at the onset of an intensity step. It must also be non-zero at points along an intensity ramp. The common first-order derivative-based methods are Roberts, Sobel and Prewitt.

7.1.2.1. Roberts Edge Method
Roberts Edge Method is a traditional method and it uses a 2x2 convolution mask to identify the rapid change in intensity values. This method is more sensitive to the noise. Roberts' edge detection is more accessible, faster to compute, and 2D spatial gradient measurement on an image. It focuses on regions that have a high spatial frequency that corresponds to edges. The input can either be a true image as in Figure 46 or a grayscale image as in Figure 47. The output is a binary image. The binary image contains value 1 in the edge region and value 0 in the background region, as in Figure 48. Pixel values at each point in the output represent the approximate absolute magnitude of the spatial gradient of the input image at that point.

Figure 46. Input image.

Figure 47. Gray scale converted image.

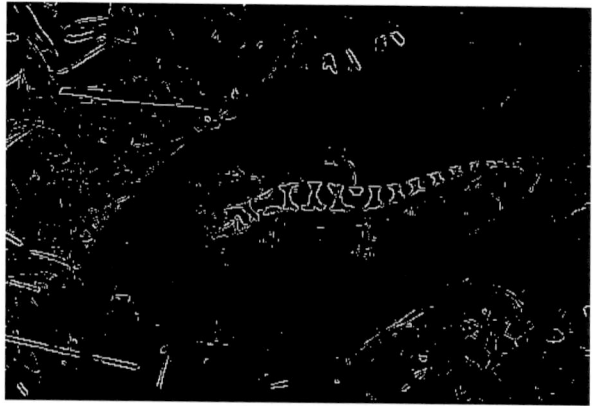

Figure 48. Output of Roberts' method.

7.1.2.2. Sobel Edge Method

The Sobel operator is not like Roberts, but it uses two 3x3 kernels to ascertain the change in intensity value. This method has less susceptibility towards noise and produces better output. Noise level is less in this method when compared to the Roberts method.

The Sobel edge detection method performs a 2-D spatial gradient measurement on an image and gives importance to regions of high spatial frequency that correspond to edges. It is generally used to find an absolute gradient magnitude at each point in an input grayscale image. The operator consists of a pair of 3x3 convolution kernels, and one kernel is just arrived by rotating the other kernel by 90°. This technique is similar to the Roberts edge method with 1 in the edge region and value 0 in the background region, as in Figure 49.

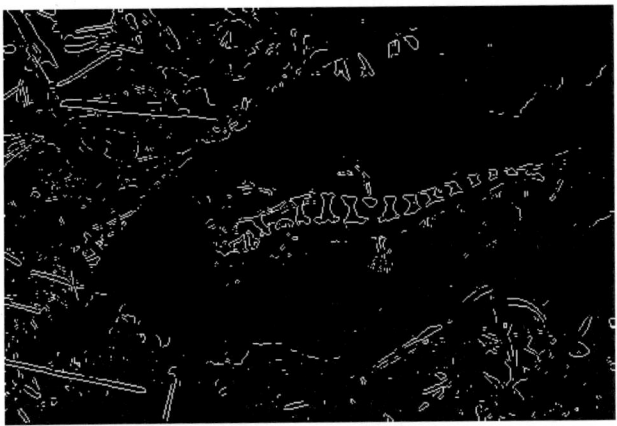

Figure 49. Output of Sobel method.

7.1.2.3. Prewitt Edge Method

Similar to Sobel the Prewitt method uses two 3x3 convolution masks and first order derivative function to identify edges. Prewitt's edge detection method is similar to Sobels' method except for the factor that image smoothing is also possible in Prewitts' edge detection method. Noise suppression is the key factor that attracts this method. Estimation of edge strength and direction is also possible. Prewitt operator is limited to 8 possible orientations. The output for this method is estimated using 3x3 neighborhoods for eight directions, as in Figure 50. All eight convolution masks are calculated. One convolution mask is then selected, which has the most extensive module.

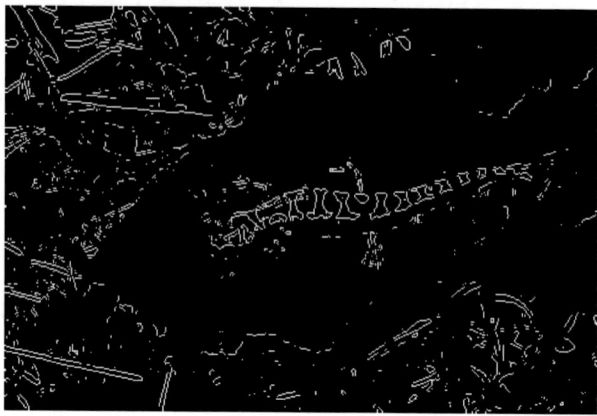

Figure 50. Output of Prewitt method.

7.1.3. Second Order Derivative Based Methods

In second-order derivatives, these functions are defined based on the criteria that it must be zero in areas of constant intensity, non-zero at the onset and an end of an intensity step. It must also be zero along intensity ramps. Unlike first-order derivative where the edges are thick, the edges are thin lines in this method. These methods are susceptible to the noise. Some commonly used methods are Laplacian of Gaussian (LoG), zero cross, and canny (Engelbrecht, 2007).

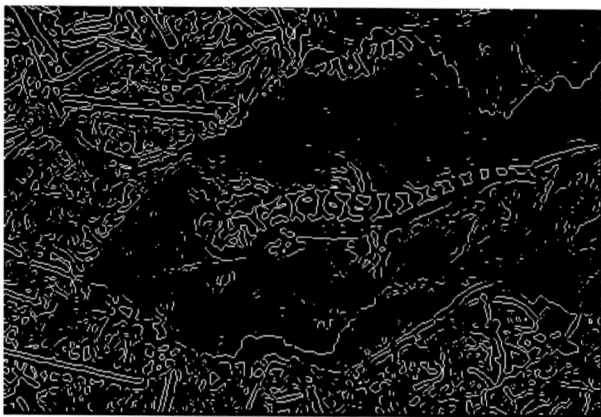

Figure 51. Output of LoG method.

7.1.3.1. LoG Based Edge Method

LoG Based Edge Method uses Laplacian and Gaussian filters to identify an image's intensity change in pixel values. The blurring effect is reduced to a great extent in this method. LoG edge detection method is used in general to filter the image using second-order derivatives. Gaussian filter is used to smooth the input image to obtain output as in Figure 51. The Laplacian operator provides an image with zero crossings to identify the location of edges. Deblurring of the image is done using the standard deviation.

7.1.3.2. Zero Crossing Edge Method

Zero crossing edge method is similar to that of LoG method except for it uses different convolution masks with different filters, as in Figure 52.

Figure 52. Output of zero cross method.

7.1.3.3. Canny Edge Detection Method

Canny is the most widely and popularly used edge detection method. This algorithm uses Gaussian to smooth the image using two 1-D Gaussian in the x and y direction. Then gradient of the image is calculated to find the change in intensity in the two directions of x and y. Edges are identified at points with a maximum gradient value and other pixel values are suppressed. The Hysteresis thresholding method is used as it uses both high and low threshold values to detect the edges. This method can identify even small lines and points in an image and is more sensitive toward noise, as in Figure 53. The main disadvantage of this method is it produces double edges and dislocated edges.

Figure 53. Output of canny method.

7.2. Region-Based Segmentation Methods

Region-based segmentation is based on the similar properties of pixel values such as intensity, color and texture. Partitions done based on these properties classify the different objects in an image. Some pixel properties assumptions in the region growing like pixels distributed within a region satisfy Gaussian distribution and mean intensity for each partitioned region are dissimilar. It is possible in this method to detect boundaries without performing any pre-smoothing process in the input image. It is also possible to segment edgeless images in this method. Selection or fixing of seed points for partitioning is a critical step in this method. Improper selection may lead to segmenting either smaller or larger regions than the required region. Thereby, this leads to over or under-segmentation. This region-based segmentation is classified into two methods: region growing and region split and merging.

7.2.1. Region Growing Method

Region growing is a method in which a region grows from a single seed point or a set of seed points as in Figure 54. A seed point is a starting point from which a region starts growing; as neighboring related pixels are added, the region is appended. When no pixels are related to the seed point and satisfied then region growth is stopped. The resultant region is the output of the region growing method in partitioning the region of interest. The success of the region growing method is purely based on the selection of seed points.

Improper selection of seed points might lead to the false classification of regions.

Figure 54. Output of region growing method.

7.2.2. Split and Merge Method

Split and merge is an alternate method for region growing in which the image is subdivided into random, disjoint sets. These regions are merged to satisfy some fixed conditions. They are repeatedly split and merged until satisfying the predefined criteria. In this method, image is subdivided into quadrants, and then quadrants are divided into sub-quadrants. These splitting are represented in the form of a quad tree with each node having 4 siblings and root as an entire image. The merging process starts when the images are partitioned and there is no possibility to partition the quadrant further. Adjacent regions are merged until there is no region left unmerged. Euclidean distance measure can be used to merge the split clusters. Segmentation done by this method has a better result as this method is simple and easier to perform.

7.3. Thresholding-Based Segmentation Methods

An important and vital method in image segmentation is thresholding. It is used to identify the boundaries in an image. The basic criteria for thresholding is that it needs a region homogeneity in intensity values and a background with varying intensity levels. The thresholding values are selected using the histogram. Suppose a histogram has a single peak value. In that case, the histogram is considered unimodal histogram, and this peak value is considered the threshold value. If a histogram has two peak values, then it is considered bimodal histograms and has two peak values to be considered for thresholding. It considers either global or local information. Thresholding performed using global information is termed contextual method, and Thresholding performed using local information is termed as non-contextual methods.

Figure 55. Output of global thresholding method.

Figure 56. Output of adaptive thresholding method.

Global thresholding, multi-level thresholding and adaptive thresholding are the major types of thresholding. Global thresholding is suitable for images with a unimodal kind of histogram. It has a single threshold value. The pixels with an intensity values greater than the threshold are considered the foreground object, and the rest of the region is considered to be the image background, as in Figure 55. If an image has more than two regions or objects with different intensity levels, then there is a need for multiple thresholding. Images which produce bimodal histograms are suitable for multiple thresholding, as in Figure 57. It has two threshold values to segment the objects in an image. The adaptive thresholding method, also referred to as dynamic thresholding, is useful for images with regions overlapping over other nearby regions in an image as in Figure 56. This thresholding is done over every window sliding block. In this way, it is processed over the entire image. The threshold value is adaptive as it changes and adapts the value according to the specific sliding block of grouping objects into clusters. Objects within a given cluster have a high degree of similarity and the objects belonging to different clusters have a high degree of dissimilarity.

Figure 57. Output of multilevel thresholding.

7.4. Clustering Based Segmentation

Clustering-based segmentation is an unsupervised method for classifying images into different clusters based on pixel characteristics. The characteristic of pixel determines the number of clusters that can be formed in an image.

Similarity and dissimilarity are the main characteristic features of a pixel in an image. Based on these features, images are clustered into groups. Each cluster is unique from the others but pixels within the clusters are similar. A similarity measure for a cluster is decided based on the application and data set. Examples of similarity measures are intensity value of pixel and connection regions of the pixel with its neighborhood. Hard means and Fuzzy means are popular clustering methods used for segmenting images. Hard means clustering is a traditional method used in the unsupervised classification of objects into groups. In an image, pixels are considered objects and grouped into clusters. According to the selection of a number of clusters, the cluster groups are formed in an image. Different cluster label index can be formed for an image based on the number of clusters, as in Figure 58.

Figure 58. Output of K-means clustering method.

Clustering algorithm generally belongs to two groups of classifications. They are as follows:

7.4.1. Partitional Clustering Algorithms

This type of algorithm attempts to directly decompose the dataset into a set of disjoint clusters based on the similarity between objects (e.g., K-Means and K-Mediods).

7.4.2. Hierarchical Clustering Algorithms

This type of algorithm decomposes the database into several levels of nested partitions, represented by dendrogram (e.g. Single Link and Average Link). The nature of Clusters can be either Hard or Soft. In Hard clustering, each object is assigned to precisely only one cluster. It is generally based on classical set theory. The clusters are disjoint in this group. K-Means (KM) and its variants belong to this category. In soft clustering, the object belongs to more than one cluster. It is based on fuzzy sets or interval sets. Clusters are overlapped in this group. Fuzzy C-Means (FCM) and Rough K-Means (RKM) belong to this category.

7.4.2.1. K-Means Algorithm
K-means is a conventional algorithm randomly selecting K objects as initial centroids and assigning remaining objects to their closer centroid based on Euclidean distance. It recalculates centroid by averaging the data point of the same cluster. These cluster steps are repeated until union.

7.4.2.2. Fuzzy C-Means Algorithm
It is one of the fuzzy clustering algorithms with soft nature. It smooths the hard nature of the K-Means algorithm. Each object is assigned to multiple clusters with a certain degree of membership. The value of membership lies between 0 and 1. Sum of the membership value of each object is 1. A larger membership value indicates higher confidence to the cluster. The clustering method is used to obtain an image of different intensity regions based on minimum distance to examine each pixel in the image and then to assign it to one of the image clusters.

7.5. Color Image Segmentation

Image segmentation is not constrained to monochrome images but can also be segmented directly on color images. In recent years, color image segmentation has gained importance due to its application in various sectors like medical, GIS, remote sensing, etc. It performs a challenging task of segmenting objects in an image similar to that of human brains capable of distinguishing millions of color from its perception. The RGB color model is generally suggested for performing the segmentation task due to its simplicity and processing speed. Each pixel is a triplet value of red, green and blue color component in a color image. The red, Green and Blue color component ratio is an important factor for color segmentation. A threshold value is set based on the intensity value of color to segment the image because intensity values are constant in similar color regions. Image segmentation is performed directly on the RGB vector space using the Euclidean similarity measure.

Objects with a specific color range can also be detected from the entire image. To do so, the required color's mean value is calculated and compared with each RGB pixel value of the image. This comparison results in getting the pixel regions within this mean value. Thus required color region can be obtained. For comparing the mean value and each RGB pixel value, Euclidean or Mahalanobis distance measure is used. When Euclidean distance measure is used for comparison, the mean value is selected as a parameter for measurement. When Mahalanobis distance measure is used as a method, mean value and co- variance matrix of the required color region are selected as a parameter for measurement. Threshold value selection is also important as a result varies for different threshold values. This type of segmentation will be advantageous when there is a need to segment a specified color object from an entire region, as in Figure 59.

Figure 59. Output of color image segmentation method.

7.6. PDE-Based Image Segmentation Methods

Due to the rapid increase in the application of image processing in numerous disciplines, it has become an essential requirement to automate the process of segmentation completely. Completely automated segmentation is used to locate the different objects in an image without manual intervention. It also identifies the shape and size of objects in an image. Methods like fuzzy logic, artificial neural networks and so on are used for the purpose of automatic segmentation. But these methods face problems like non-convergence, to stuck in local minima and not being more effective towards the noisy image. PDE-based segmentation overcomes this problem and produces optimal segmentation output. The popular automatic segmentation in PDE-based segmentation is the active contour model and geodesic active contour model. The active contour model is a parametric method used to identify the boundary of an image using edge information. The geodesic active contour model uses both the region and edge information from an image to identify the boundary of an image. Snakes are propagated using methods like gradient vector flow, level set theory, fast marching method and active geometric contours. In this section different active contour models are discussed in detail.

7.6.1. Active Contour Model

An active contour model is a deformable model or snake used to grow or evolve a curve to segment an image. This method is very much useful for real-time images and is also very much effective for both 2D and 3D images. Snake sets a constraint to follow a minimal energy path. This makes the snake to grow towards the image boundary. Snake uses a concept similar to that of the region growing approach, as in Figure 60. As a primary step, snake contour takes the initial seed pixel from the region of interest. Snakes' curve points are generally represented using B-Spline curves along with specific control points. As it is parametrized, the control points are described in the parametric form. To make the snake move, it needs a force or energy to move along boundaries. These force or energy functions can be calculated from the image itself and are used to control the snake's movement. Internal and external energy are the two types of snakes' energy. Internal energy is based on the intrinsic properties of an image, like boundary length or curvature. External energy is based on the structure of an image. It is used to control the association between the snake and the image. Therefore the total energy for the snake to move is determined by combining all the factors like boundary length, curvature and the image itself. To reach the optimal solution these parameters can also be adjusted. This iteration is continued until the snake model tightly closes the object's boundary in an image.

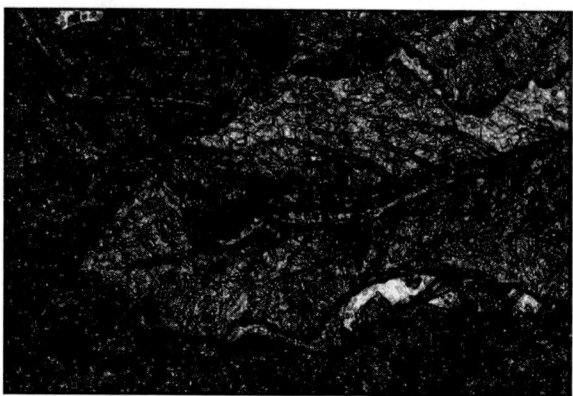

Figure 60. Output of active contour method.

7.6.2. Geodesic Active Contour Model

The geodesic active contour model is also referred to as the geometric active contour model. It is considered an alternative to the snakes model. In this method similar to region split and merge, the object's contour splits and merges to identify the boundary of an object as in Figure 61. Energy is minimized for the curve by applying an objective function obtained by solving the Euler-Lagrange equation. The zero level set of a function is used to calculate this geodesic distance. Unlike the active contour model, this method is parameter-free and the changes in geometrical properties are also handled easily. Due to the use of level set theory and the easy handling of geometric properties, this method comparably had a better outcome.

Figure 61. Output of geodesic contour method.

7.7. Hybrid Segmentation Method

Edge-based and region-based methods are the most commonly used methods of image segmentation. Still, they mostly encounter problems separating the required region of interest from an image. The major problem faced in edge-based segmentation is the selection of threshold value. Improper selection of threshold value for binary conversion of image produces disconnected edges that lead to under-segmentation of image. In region-based segmentation, the major problem is the selection of appropriate seed point pixels to obtain the region of interest. Improper selection of seed points may lead to over-segmentation of the image.

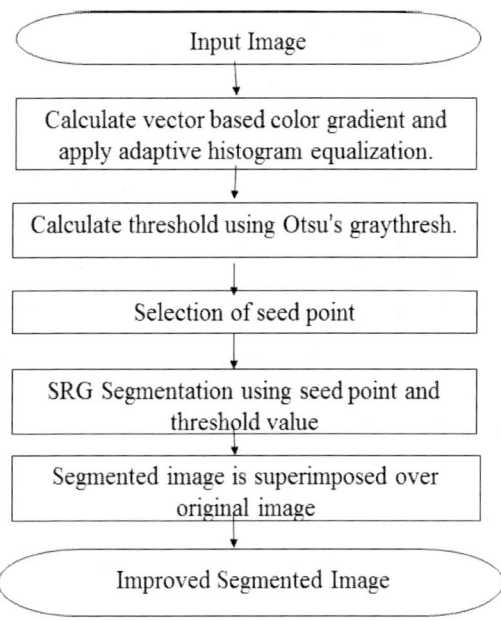

Figure 62. Steps involved in the hybrid method.

So to overcome these drawbacks, hybrid segmentation is proposed by combining edge-based and region-based segmentation techniques. Vector-based color gradient is calculated for obtaining the edges of the image and adaptive histogram equalization is applied to it. Histogram equalized image is then taken as input for region growing segmentation method. Threshold value selection and seed point selection are two important tasks in the region growing segmentation method. Otsus' threshold is used to set the threshold value for SRG using the grayscale image to fix the threshold value for the region growing method. The steps involved in developing the hybrid segmentation method are shown in Figure 62. This hybrid method improves the segmentation accuracy by combining the edge-based and region-based segmentation techniques.

7.7.1. Edge Detection Using Vector Based Color Gradient

As edge detection is an important module in image segmentation, it is computed directly on the RGB color space model. In monochrome images, the scalar function is used for gradient calculation. In color images, the vector function is used for gradient calculation. The gradient, in general, is defined as a two-dimensional function with a significant property of identifying the greatest magnitude direction in a function. The vector vector-based color gradient magnitude is used to compute the first-order derivatives. The first-order derivatives are used for deriving the edges in an image. The square root of the largest Eigen value in a vector-based color gradient is considered its magnitude value. A gradient image is derived using the magnitude and direction of the gradient. Another way to obtain the edges is to manipulate the gradient in individual color components (red, green and blue) and then add these individual color component gradients into a single color gradient image. The output of these two approaches might not be similar and there might be slight differences. The -vector-based color gradient has a higher accuracy and reliability than the combining of individual color component. The vector-based color gradient is applied to the input image and then adaptive histogram equalization is applied to the color gradient to enhance the image.

7.7.2. Region Growing Method

In region-based segmentation, the images are partitioned into various regions. This technique assumes that adjacent pixels in the same region have similar features like grey level, color value, or texture. This region-based segmentation is classified into two methods: region growing and region split and merging. The region growing method integrates sub-regions of pixels into larger regions based on predefined growth conditions. The selection of a set of seeds is a complicated job. From this seed grows regions appending with other seeds which are similar. However, SRG suffers from an automatic selection of initial seeds to attain more accurate segmented images. The seed value can be either a single seed value or an array. If the seed value is an array, it contains the value 1 at the coordinates of similar seed points and the value zero in the rest of the region. If the seed value is scalar, then the seed value is either 0 or 1. The seed value is selected on the basis of the edges of the gradient image.

The threshold value can also be either an array with a similar image size or a single global threshold value ranging from 0 to 1. The threshold value is calculated using Otsus automatic threshold selection method. The threshold value selected by the Otsus threshold method and the seed point determined manually is used to segment the histogram equalized color gradient image based on the region growing method. The selected seed point is grown towards the entire image with a threshold value as its criterion. The resultant segmentation method can detect edges with greater accuracy, as in Figure 63.

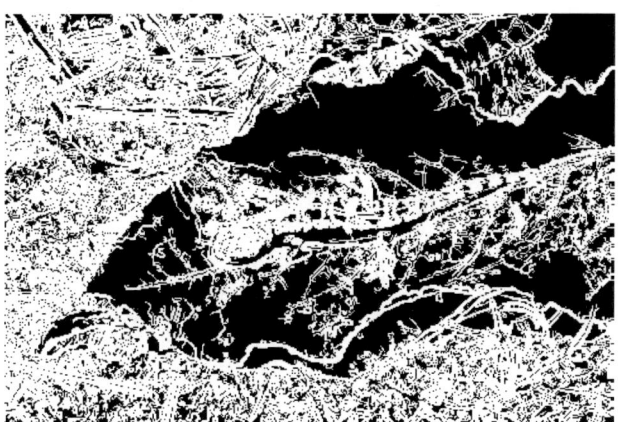

Figure 63. Output of hybrid segmentation method.

7.8. Optimization-Based Edge Detection

Optimization-based edge detection methods produce an exact edge map with exact edge information using constrained and unconstrained optimization. It overcomes problem like false edges, missing edge regions, discontinuity in edges and noises. The constrained-based method uses the concept of constrained optimization to focus on the problem of edge detection as an optimization problem. Constrained optimization helps to produce an optimized edge map with higher accuracy. Edges are continuous, long, and definite with no false edge regions. Mathematically, Constraint optimization is used for solving optimization problems that are restricted by constraints in the objective functions. Penalty method, Lagrange multiplier, and augmented Lagrange multiplier are the common nonlinear constrained optimization used to solve optimization problems as listed in Figure 64. The penalty method was used in the many edge detection method as it is flexible to both the equality and inequality constraints. In unconstrained-based optimization, it uses the unconstrained non-linear derivative-based optimization technique that views edge detection as a minimization problem of the objective function. Steepest descent, conjugate gradient and newtons' method are the non-linear derivative-based optimization methods that efficiently search through iterations until an optimal solution point is reached for the edge regions.

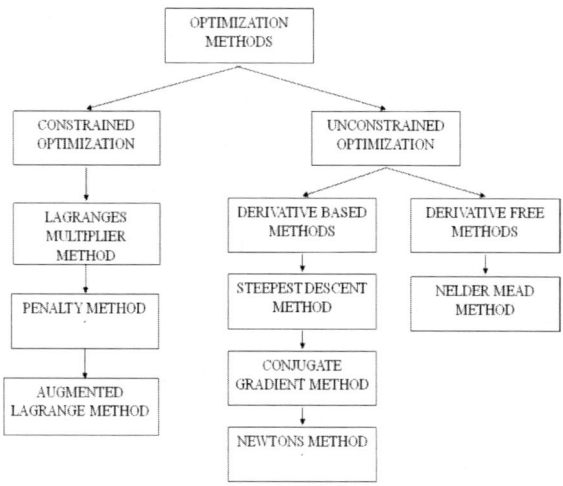

Figure 64. Different types of optimization.

7.8.1. Improved Edge Detection Method Using Non-Linear Constrained Optimization

Constrained Optimization is used for finding the optimal point either as minimization or maximization in an objective function. It is influenced by two important factors called criteria and constraints. Criterion expresses the purpose of optimization either as minimization or maximization of the objective function. Constraints are the restrictions imposed over the input variables of the objective function that express the relationship among these input variables in mathematical terms. Constraints can be either equality or inequality constraints. Constrained optimization is formulated mathematically as,

$$\min_{x \in R^n} f(x) \quad subject\ to \quad \begin{cases} c_a(x) = 0, a \in E \\ d_a(x) \geq 0, a \in I \end{cases}$$

where $f(x)$, $c(x)$, $d(x)$ are the real-valued functions on the subset of R^n (Real value numbers), $f(x)$ is an objective function for the minimization problem, $c(x)$ are set of equality constraints and $d(x)$ is set of inequality constraints, E and I are set of values for c and d with a as the number of constraints.

An optimization problem with constrained variables is solved by converting the problem into an unconstrained optimization. But the algorithmic approach of the unconstrained method is not well suited for these constrained methods after conversion as it causes the Maratos effect, i.e., it causes a lack of convergence. Therefore, non-linear methods like penalty, Lagrange multiplier, and augmented Lagrange multiplier, non-linear methods like penalty, Lagrange, and augmented Lagrange multiplier methods are used in constrained optimization to attain optimal result with faster convergence rate. The penalty method is very much of use to solve constrained problems with equality or inequality constraints, and the Lagrange method is concerned only with problems related to equality constraints. The augmented Lagrange method is similar to that of penalty method. Still, the major difference is that it adds constant term to the unconstrained objective function that is similar to that of Lagrange multiplier. Still, it is not similar to the Lagrange multiplier method.

7.8.1.1. Penalty Method

The penalty method usually converts the constrained problem into the unconstrained problem by adding a penalty function to the objective function for ideal convergence. The penalty method with equality constraints has a value equal to zero. The penalty method with inequality constraints have a constraint value less than or equal to zero for the minimisation problem and has constraint value greater than or equal to zero for the maximization problem. The penalty function uses a penalty multiplier to convert the problem into an unconstrained method. Penalty multiplier uses a penalty parameter, a scalar value multiplied along with the constraint to obtain a new objective function. After deriving the new objective function next step is to optimize the new objective function using derivative-free unconstrained optimization techniques. Nelder-Mead, also known as the direct or simplex search method, is applicable for optimization problems were the derivative method is complex, expensive and not applicable. This method is also useful when the function is discontinuous and noisy. It requires only function evaluation and not derivative values for optimization. To solve a minimization problem, f is considered a real-valued function with simplex S, a polyhedral set with n+ 1 vertex a1, a2, a3, an+1. An associated matrix (A(S)) with linear independence in Rn (set of real numbers) is

$$A(S) = [a_2 - a_1, a_3 - a_1, \ldots \ldots a_{n+1} - a_1]$$

The simplex is non-degenerate if A is a non-singular matrix, i.e., its vertices are not coplanar. This algorithm searches the appropriate value through iterations. It compares the function values on the vertices removes the vertex with the worst function value and then replaces it with another vertex point with a better value. The selection of a new point is based on the reflection, contraction, shrinkage and expansion of the simplex along the line that joins the worst vertex and the centroid of the other vertices. Let $f(a1), f(a2), f(a3), \quad f(an + 1)$ be an ordered vertices at any specific step of the algorithm and it can be expressed as

$$f(a_1) \leq f(a_2) \leq f(a_3) \leq \ldots \ldots \leq f(a_n + 1)$$

where $f(a)$ is a function for the constraints with a as the constraints. Thus the new points replace the worst vertex to solve the optimization problem.

7.8.2. Lagrange Multiplier Method

Lagrange multiplier is a powerful method to solve optimization with equality constraints. Instead of using a penalty function like the penalty method, Lagrangian function is used to solve the optimization problem. The Lagrangian function uses a Lagrange multiplier to convert the constrained problem into an unconstrained problem. A Lagrange multiplier is an unknown scalar value combined with the objective function to form a new objective function. The number of Lagrange multipliers required for a function is determined by the number of equality constraints found in the function. If Z equality constraints exist, then the Lagrange multiplier required to solve the problem is also Z.

Let the constrained optimization problem subject to the equality constraint be expressed as

$$min\ f(x)\ subject\ to\ C_i(x) = 0, x = (x_1, . x_n)^T$$

where f(x) is the objective function, C is the constraint, x is the set of values in the vector and n is the number of elements in x. Next, to obtain a new objective function called the Lagrangian function, the Lagrange multiplier is added to the existing objective function as

$$L(x, \lambda) = f(x) + \sum_{i=1}^{z} \lambda_i C_i(x)$$

where λ is a scalar value of the Lagrange multiplier is added to the equation. Finally, to solve the minimization problem, partial derivatives of the new objective function ($L(x, \lambda)$) are set to zero as,

$$\frac{\partial}{\partial x} L(x, \lambda) = \nabla f(x) + \sum_{i=1}^{z} \lambda_i \nabla C_i(x) = 0$$

$$\frac{\partial}{\partial x} L(x, \lambda) = C(x) = 0$$

This method produces an optimal solution for problems with equality constraints.

7.8.3. Unconstrained Non-Linear Optimization Technique Based Enhanced Edge Detection Method

Despite the availability of numerous edge detection methods, more algorithms are developed for edge detection. The reason for this continued exploration is that outputs of most edge detection techniques face problems like double edges, false edges, edge displacement, discontinuity in edges, missing true edges, noises and edge over-detection. The concept of unconstrained non-linear derivative-based optimization technique can overcome the above-mentioned deficiencies. Unconstrained non-linear programming is a model that works on continuous decision variables similar to linear programming but has non-linear constraints and non-linear objective functions. The objective function can be of either a single variable or multivariable function. Unconstrained non-linear methods are more advantageous than the linear methods as they have more probability of finite optimal solution. Another favorable point is that most of the non-linear methods are constraint-free. If few simple constraints are found in the problem, it can also be neglected to attain a feasible solution.

Standard non-linear optimization methods are golden search method, quadratic approximation method, Nelder-Mead method, steepest descent method, conjugate gradient methods and newton's method. Golden search and Quadratic approximation method are one-dimensional search methods for minimization of a function over a single variable. The golden search method uses golden section search, a unimodal objective function that rapidly constricts the interval value to have an optimum. The quadratic approximation is a three-point bracketing pattern for unimodal objective function containing three decision variables with objective value at its minimum or maximum to reach the optimum. The Nelder-Mead is a non-derivative search method to minimize objective function over multi variable values.

Steepest descent method, conjugate gradient methods and Newton-Raphson method are derivative-based search methods that substantially accelerate the search progress s to search for local minima. The steepest descent method is used to search with the step size to find the minimization of an objective function for N variables. Step size is the number of iterations determined based on the distance to be moved in the gradient vector. It searches toward the greatest decrease (negative gradient vector) to reach the local minimum point. The drawback of this method is the zig-zag approach used to reach the minimum threshold. The Newton-Raphson method finds the minimization of N variables with the objective function using the gradient vector. This method uses the first- and second-degree polynomials of Taylors' approximation for higher efficiency. The conjugate gradient method is similar to the steepest descent and Newton-Raphson method as it also uses the gradient vector to find the minimum value of an objective function. The drawback of this method is the requirement of more computing effort to reach the minimum point. In this chapter, Newton-Raphson method was used to find the better edge points in an image as the convergence rate is fast in this method when compared with other derivative-based non-linear methods. Implementation of this method for edge detection techniques has solved the problem of finding true edges. This method can also distinguish between local noise and image structure variations.

7.8.3.1. Newton-Raphson Based Segmentation Method

The steps involved in the development of Newton-Raphson based segmentation are given below;

Step 1. Select an input image.
Step 2. Compute vector-based color gradient directly on the RGB image.
Step 3. Calculate the standard deviation value from the gradient image to fix a threshold value for edge localization.
Step 4. Initialize starting pixel point in the gradient image.
Step 5. Selection of pixels based on a threshold condition. If the threshold value is satisfied, proceed further else, consider the next pixel point.
Step 6. Derive a hypothetical objective function.
Step 7. Iterate the objective function with the pixel value selected in Step 5 to reach the local optimum point.
Step 8. Repeat Steps 5 to 7 for all pixel values.
Step 9. Suppress non-edge pixels using kernel operation.
Step 10. An improved edge detection output is achieved.

7.8.3.2. Vector Based Color Gradient Calculation

Color gradient method takes an advantage over the scalar gradient method as it works efficiently in noise affected image and low contrast image. Vector based color gradient is different from the scalar space gradient method as it applies the concept of vector space instead of scalar space. The vector space-based gradient is implemented in various ways. In the scalar function, the gradient is the maximum rate of change of the function at the pixel coordinate points in the image. This concept is applied over the unit vectors of individual color components of the RGB color space. An image I(m,n) is considered as a vector function comprising three color component axis A(m,n), B(m,n) and C(m,n) and m,n are considered as the pixel points for the three components. The horizontal and vertical direction gradients(x, y) are defined using the formula,

$$X = \frac{\partial A}{\partial m}a + \frac{\partial B}{\partial m}b + \frac{\partial C}{\partial m}c$$
$$Y = \frac{\partial A}{\partial n}a + \frac{\partial B}{\partial n}b + \frac{\partial C}{\partial n}c$$

Next the dot product for these vectors are calculated from equation 1 & 2 as mentioned below

$$K = X \cdot X = X^T X = \left|\frac{\partial A}{\partial m}\right|^2 + \left|\frac{\partial B}{\partial m}\right|^2 + \left|\frac{\partial C}{\partial m}\right|^2$$

$$L = Y \cdot Y = Y^T Y = \left|\frac{\partial A}{\partial n}\right|^2 + \left|\frac{\partial B}{\partial n}\right|^2 + \left|\frac{\partial C}{\partial n}\right|^2$$

$$G = X \cdot Y = X^T Y = \frac{\partial A}{\partial m}\frac{\partial A}{\partial n} + \frac{\partial B}{\partial m}\frac{\partial B}{\partial n} + \frac{\partial C}{\partial n}\frac{\partial C}{\partial m}$$

The maximum rate of change of the function I(m,n) is given by

$$\theta(m, n) = \frac{1}{2} \tan^{-1}\left[\frac{2G}{K-L}\right]$$

The rate of change of I(m,n) in the direction of $\theta(m,n)$ is calculated using

$$D_\theta(m, n) = \left\{\frac{1}{2}[(K+L) + (K-L)\cos 2\theta\,(m,n) + 2G \sin 2\theta\,(m,n)]\right\}^{\frac{1}{2}}$$

As a final step, the magnitude of I(m,n) is calculated using

$$V = \sqrt{(D_\theta)}$$

The final value, 'V' is the vector-based color gradient image for the given input RGB image.

7.8.3.3. Finding Edge Points Using Newton-Raphson Method

The Newton-Raphson method extends the quadratic approximation approach of processing functions with multiple independent variables. If an appropriate objective function is chosen, then the convergence of the objective function to an optimal solution is definite. Unlike the steepest descent method, this method uses the second-order Taylor approximation and the first-order Taylor approximation. First order approximation of Taylor polynomial is linear in approach over the directional components of gradient, whereas second-order approximation is non-linear in approach. First derivatives or gradients give detailed information about objective function changes near a current solution. It explains the slope or rate of change with small variations in the objective function. Second derivatives or Hessian matrix provide information about the function's nature in the neighborhood of a current solution. Second derivatives are used for multi-variable objective optimization to find the optimum of the gradient function.

The Taylor polynomial of degree two to minimize the function, f(a) is defined as

$$f(a) \cong f(a_i) + g_i^T[a - a_i] + \frac{1}{2}[a - a_i]^T h_i[a - a_i]$$

where 'i' is the iteration value, f(a) is an objective function with respect to the independent variable 'a', i.e., the intensity value at each pixel coordinate point in the image and 'g_i' is the gradient vector and 'h_i' is the Hessian matrix.

The first derivative or gradient is defined as

$$g_i = \frac{d}{da}f(a) = \begin{bmatrix} \frac{\partial f}{\partial a_1} \\ \frac{\partial f}{\partial a_2} \end{bmatrix}$$

The second derivative or Hessian matrix is defined as

$$h_i = \frac{d^2}{da^2}f(a) = \begin{bmatrix} \frac{\partial^2 f}{\partial a_1^2} & \frac{\partial^2 f}{\partial a_1 \partial a_2} \\ \frac{\partial^2 f}{\partial a_2 \partial a_1} & \frac{\partial^2 f}{\partial a_2^2} \end{bmatrix}$$

Finally, from the above equations, newton's method is defined as

$$a_{i+1} = a_i - \frac{f(a_i)}{f'(a_i)}$$

The intensity value of the pixels '(a_i)' of the vector-based color gradient was considered for input to the objective function 'f(a)' to determine the optimal edge region. The objective or fitness function for pixel value (a) manipulation is defined in this proposed work as follows:

$$f(a) = 4a^2 - a - 1$$

This equation is iterated at each pixel position until the final pixel position was reached. Thresholding technique was used to select the pixels for optimization, and standard deviation value of the gradient image was calculated to fix the threshold value. Pixel values that satisfy the threshold value were iterated through the fitness function until convergence is reached and other pixels are not considered for iteration. In such a manner, pixels related to edge region are taken one at a time and processed through Newton-Raphson iteration. The resultant image after applying this equation forms the entire edge map. As the last step, the kernel was used to manipulate over neighborhood pixel region to suppress the unwanted edge regions that are marked. The edge output achieved is long and continuous and its accuracy is better when compared with other derivative-based optimization methods and existing methods for edge detection.

7.9. Soft Computing Edge Detection Techniques

In contrast to conventional hard computing techniques, soft computing is suitable for data that are not crisp, definite and certain. It inherits the reasoning and learning ability of the human mind and is inspired by the characteristics of the biological system. Fuzzy logic, neural network and genetic algorithm are the main soft computing techniques used to solve problems in machine learning and artificial intelligence. These methods are useful for the situation which re9uires learning, training, classification, inference, prediction and optimization. These methods are not competitive but complementary as each method has its own methodologies and design model. Neural network is adaptive in nature and is used to solve problems using self-learning algorithms. Genetic algorithm is an evolutionary computing technique used to search for an optimal solution for a problem from the available search space. Fuzzy logic is used to solve a problem related to fuzzy sets using if-then rules. Performing hybridization among soft computing methods paves the way for a successive effect. Genetic algorithm and fuzzy logic have been hybridized sequentially to achieve an efficient edge detection method capable of producing an enhanced and continuous edge without any discontinuity issues (Castillo et al. 2012).

7.9.1. Fuzzy Logic and Genetic Algorithm-Based Edge Detection

The steps involved in developing the algorithm are given in Figure 65.

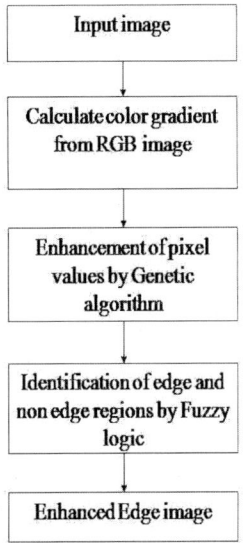

Figure 65. Steps followed to develop genetic-fuzzy hybrid edge detection algorithm.

Figure 66. Input original image.

Calculating Color Gradient: The input image in RGB color space is used to calculate the color gradient as in Figure 66. The color gradient method calculates the gradient value in each color component (red, green and blue components) of an image using vector space. It then combines them to calculate the gradient for the entire image. Calculating the 2-D gradient for each component of an image f(x,y) is done using the formula,

$$\nabla f \equiv grad(f) = |g_x| + |g_y|$$

where, f is an individual component of an image and x and y are the coordinate values of f, g_x and g_y are the gradient value in x and y direction. This method can produce a gradient image with low contrast than the scalar space-based gradient calculation (Rajasekaran & Pai, 2003).

Table 2. Algorithmic steps involved in the phase of genetic algorithm for enhancement

Step 1	Get Initial pixel value (x0) from gradient image.
Step 2	Set population size.
Step 3	Set number of iterations.
Step 4	Generate initial population randomly.
Step 5	Decode Initial population in binary form to decimal value.
Step 6	Evaluate the initial population by fitness function, $((x2 - 4)2/8) - 1$
Step 7	If converged terminate the process and move to Step 13, else proceed.
Step 8	Select two parents by Random selection technique to generate next generation.
Step 9	Mate two parents by shuffle crossover technique to generate new desired offspring.
Step 10	Mutate by reversing technique to avoid local minima errors and to recover from loss of genetic information.
Step 11	Move to Step 6.
Step 12	Evaluate the newly generated offspring population.
Step 13	Replace new pixel value (x1) in x0.

Enhancement of pixel values by Genetic Algorithm: The next step is to enhance the image for smoothing to reduce the noise level. Genetic algorithm (GA), a heuristic optimization approach, is applied over the gradient image. It is an evolutionary computing technique for a directed random search to reach an optimal solution. This optimization method is considered over the traditional techniques for enhancement as it reaches optimal solution and its' evaluation over the pixel value is done using the fitness function rather than functions based on derivatives. The other reasons are its flexibility, adaptability, understandability and requirement of less mathematical knowledge. The steps involved in this phase of the genetic algorithm for enhancement are explained in Table 2.

Finally, new optimized pixel values are replaced to get an enhanced output image.

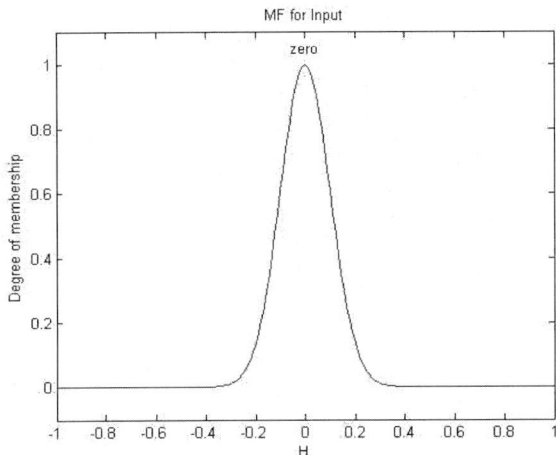

Figure 67. Membership function for the input variable.

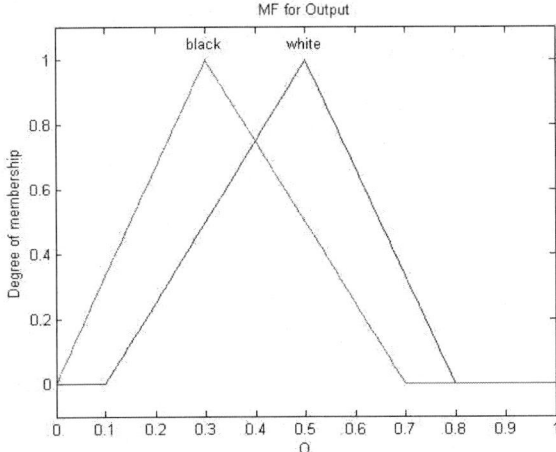

Figure 68. Membership function for the output variable.

Figure 69. Output of sequential hybridization Fuzzy-genetic method.

Identification of Edge Regions by Fuzzy Logic: Fuzzy logic is an extension of Boolean or conventional logic used to analyze partially true data. The value of fuzzy logic lies within the range of 0 to 1 in contrast to Boolean logic where the value is either 0 or 1. So, fuzzy is considered a simplified technique of the conventional method. The main advantage of this method is its flexibility in bringing together detailed information about the system. This method is useful for identifying the exact edge regions in an image as it has a higher ability to manage vagueness and ambiguity information in an image. Mamdani Fuzzy Inference System is implemented to extract the edge regions in an image. It performs the input to output mapping operation using fuzzification and defuzzification processes.

Fuzzification is a process in which the crisp set of input is converted into the fuzzy set and are represented using a membership function. Defuzzification is a process in which the output in the fuzzy set is converted back into a crisp set. So, using fuzzy logic, the edges regions are identified exactly and provide an exact edge map of the image. The output of an enhanced color gradient image using a genetic algorithm is taken as input for fuzzy logic. Steps involved in the phase of fuzzy logic for identifying the edges are described in Table 3.

Table 3. Algorithmic steps involved in the phase of fuzzy logic for identifying the edges

Step 1	Consider GA applied output as input (I) for fuzzification.
Step 2	Set Gaussian membership function for Input variable (I) as in Figure 67.
Step 3	Set Triangular membership function for output variable (O) as in Figure 68.
Step 4	Set a Fuzzy rule to differentiate edge and non-edge pixel regions.
Step 5	Evaluate Fuzzy rule and identify its consequences.
Step 6	Combine the Consequents to get a final output.
Step 7	Defuzzify the fuzzy result into crisp values using centroid method.
Step 8	Enhanced edge image is achieved

7.10. Metaheuristic Algorithms in Edge Detection

Alternative to conventional methods are the metaheuristic optimization techniques suitable for problems that are not deterministic and have no acceptable solution. The term metaheuristic has derived from the Latin words meta and heuristics, as in Figure 70. It is stochastic and is suitable for multi objective and combinatorial optimization problems. Unlike conventional methods, these metaheuristic techniques are mostly nature-inspired and these algorithms are developed using several parameters relevant to the problem and technique. It is useful for finding optimal global solutions for a problem with less computational complexity (Baghel et al. 2012; Fister Jr et al. 2013).

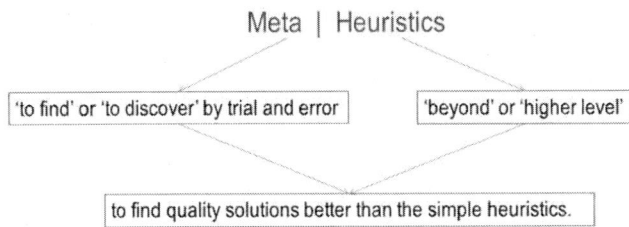

Figure 70. Meaning of the term 'Metaheuristics'

Exploration and exploitation are the two major components of metaheuristic methods, as in Figure 71. Exploration is the diversified search for the solution in the entire search space. Exploitation is the intensified search for the solution in a specific search space. It must be combined in the right proportion to reach a globally optimal solution. Metaheuristic algorithms are classified into two categories as trajectory method and population-based method (Binitha & Sathya, 2012).

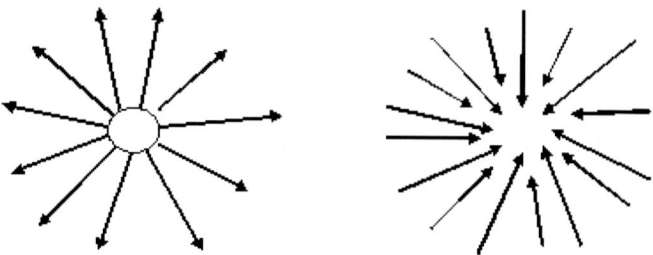

Figure 71. Diagrammatic representation of exploration and exploitation components.

Metaheuristic algorithms based on a single solution are termed Trajectory methods. Examples of these methods are Tabu search, simulated annealing, GRASP method and guided local search method. Metaheuristic algorithms based on a set of solutions are termed population-based methods. Some examples of these methods are genetic algorithms, particle swarm optimization, ant colony optimization and so on (Gandomi & Alavi, 2012; Chen et al. 2014).

This section discusses metaheuristic algorithms like particle swarm optimization, ant colony optimization, genetic algorithm and an efficient edge detection using flower pollination algorithm.

These algorithms have a set of solutions explicitly to work on to obtain an exact optimal solution for a problem. Evolutionary Computation and Swarm Intelligence are the two major artificial intelligence techniques contributing to population-based methods. Darwins' evolutionary theory inspired the evolutionary computing paradigm and it mimics the natural evolution process. Genetic algorithms, genetic programming, Differential evolution and strategies are some of the popular evolutionary computation methods (Prakash & Shetty, 2015). Swarm intelligence is inspired by the social behavior of organisms that live in swarms or colonies. Particle swarm optimization, ant colony optimization, bacterial foraging optimization, and artificial bee colony optimization are some of the commonly used Swarm intelligence techniques (Jordehi, 2015). In identifying the edges, the most widely used population-based metaheuristic algorithms are genetic algorithms, particle swarm optimization and ant colony optimization methods (Maia et al. 2012). These algorithms and their applications in edge detection are discussed in detail in this section:

7.10.1. Genetic Algorithms (GA)

It is a widely recognized and applied evolutionary computation algorithm for global search problems. It originates from the 1960s and was developed by John Holland at University of Michigan. It is stochastic and adaptive in nature, and survival to the fittest is the basic principle of this algorithm. The population of individuals is selected randomly and evolved through generations by iterations to produce a better solution. By iterating and evolving the population of individuals through generations, the weaker problem solution is replaced by a new optimal solution. The major advantage of this method is that it requires less mathematical knowledge. The simple structural flow of a Genetic algorithm is depicted in Figure 72. This basic and simpler concept of genetic algorithm is applied over the image pixels to obtain the exact edge regions from an image (Bhandarkar et al. 1994).

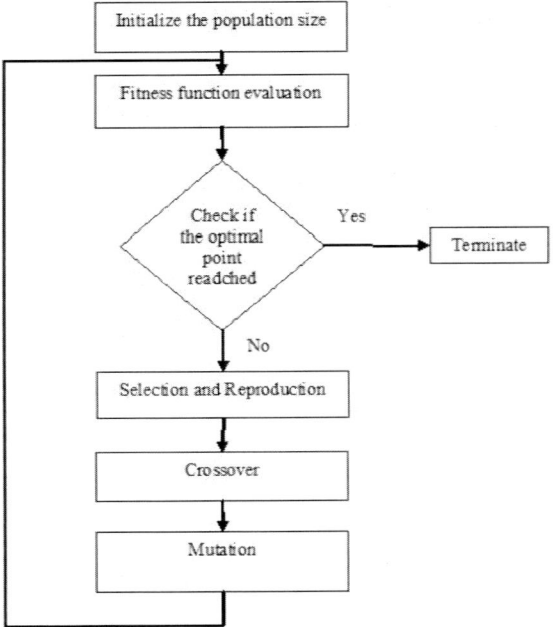

Figure 72. General flowchart of genetic algorithm.

The initial step of the genetic algorithm is to initialize the population of solution or chromosomes randomly. There are different encoding techniques available for representing the chromosomes. Some standard encoding methods are binary encoding, octal encoding, hexadecimal encoding, permutation encoding, value encoding and tree encoding. Usually, these chromosomes are represented in a bit string where each bit represents specific characteristics related to the solution. Each chromosome in the population must be evaluated for its fitness using a suitable fitness function. From the result obtained from the evaluation, the best chromosome is considered for further operations of selection, reproduction, crossover and mutation. Selection is the process of selecting two parents from the available solution population to perform the crossing operation. Commonly available selection methods are roulette wheel selection, random selection, rank selection, Boltzmann selection and tournament selection. Generally, the random selection method is used for selecting the best parents.

Next, a crossover operation is performed using these parents to produce a new offspring. Single point crossover, two point crossover, multipoint crossover, uniform crossover and matrix crossover are some available crossover techniques. Among these methods, single-point crossover is widely used in the traditional basic genetic algorithm approach. The next step is to perform the mutation process to prevent the problem from being confined to the local minima point. The mutation is done through any of these available methods like flipping, interchanging, reversing, and mutation probability. These operations are iterated until an optimal solution is reached for the problem.

7.10.2. Ant Colony Optimization (ACO)

It is a swarm intelligence-based optimization algorithm developed based on the foraging behavior of ants. The phenomenon of communicating between ants to find the shortest path for the food source from the nest is stigmergy. It is an indirect communication between the ants and also with the environment. A chemical substance called pheromone is deposited over ant movement from food source to nest and vice versa. The high thickness of the pheromone attracts or communicates to other ants to move towards that direction than the other pheromone deposit direction. This thicker deposit of pheromones is usually found in the shortest path between the food source and the nest. This concept of ant foraging behavior forms the base for ant colony optimization. This method is applied to search for an optimal solution for numerous problems and is proved to be a successful optimization algorithm. This ACO algorithm is iterated until an optimal solution is reached for a problem. The basic flowchart of an ACO algorithm is depicted in Figure 73. This simple ant colony optimization algorithm is applied over an image to identify an exact edge regions in an image (Baterina & Oppus, 2010).

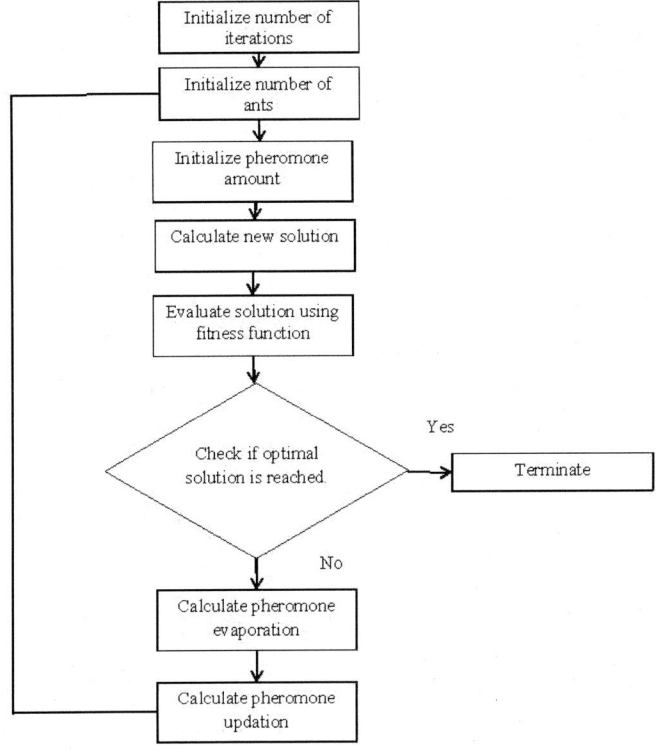

Figure 73. General flowchart of Ant Colony Optimization.

ACO is similar to the concept of graph theory of finding the shortest path between two nodes. The primary step in ACO optimization is to ze the number of ants, iteration, and initial pheromone amount and construct ant solution. All ants move from the nest to the food source in incremental steps. At each node position, each ant decides which ant to move next. Selection of the next node is made using transition probability, using the equation,

$$P_{mn}^{x}(t) = \begin{cases} \dfrac{A_{mn}^{\alpha}(t)}{\sum_{n \in N_m^x} A_{mn}^{\alpha}(t)} & if\ n \in N_m^x \\ 0 & if\ n \notin N_m^x \end{cases}$$

where 'x' represents ants and 'n_x' is the total number of ants, 'm' is the current node, 'n' is the next node, 'N_m^x' is a set of possible nodes that are connected to 'm', 'A_{mn}' is the pheromone concentration value. 'a' is a positive constant value used to strengthen the pheromone concentration. A larger value of 'a' may lead to rapid convergence and leads to a sub-optimal solution. When the ants reach the food source and return to the nest, it deposits a pheromone which is calculated using the following equation,

$$\Delta A_{mn}^x \propto \frac{1}{L^x(t)}$$

where '$L^x(t)$' represents path length constructed by ant 'x' at step 't'.

These pheromone values are updated using the following equation,

$$A_{mn}(t+1) = A_{mn}(t) + \sum_{x=1}^{n_x} \Delta A_{mn}^x(t)$$

Next, pheromones deposited in the path must be evaporated to prevent premature or earlier convergence and for more exploration by ants. Pheromone evaporation is calculated using the following equation,

$$A_{mn}(t) \leftarrow (1-V) A_{mn}(t)$$

and

$$V \in [0,1]$$

where 'V' is the rate at which the pheromone is evaporated.

If the optimal solution is reached, then the process is terminated; else, the process is iterated until an optimal required solution is reached for the problem.

7.10.3. Particle Swarm Optimization (PSO)

Particle Swarm Optimization is also a swarm intelligence-based optimization algorithm developed based on the social behavior of bird flocking and fish schooling. Equivalent to a genetic algorithm, particle swarm optimization has 'swarm' in analogy to 'population' and has 'particle' in analogy to 'individual'. Each particle in the swarm exhibits simple behavior and their collective behavior in the swarm have a habit of solving complex problems more easily. Swarms are spread over the entire hyperspectral search space. Particles change their position in a swarm using the velocity factor, which reflects the particle's knowledge and the knowledge of it's neighborhood particle. The three major velocity components are previous velocity, cognitive, and social. The cognitive component expresses particle's knowledge.

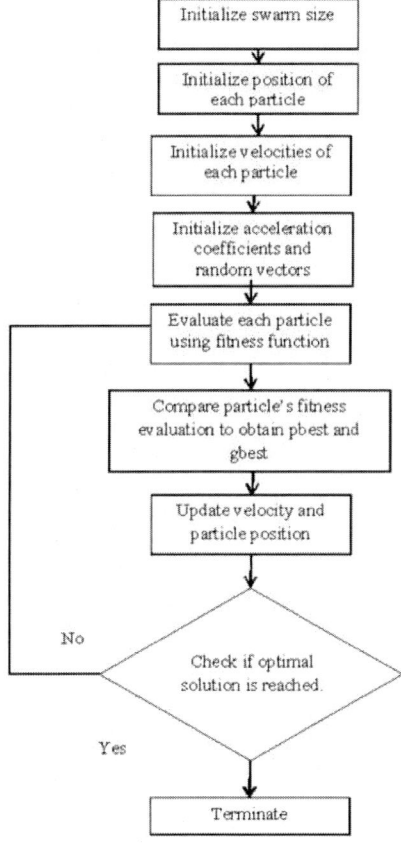

Figure 74. General flowchart of particle swarm optimization.

The social component expresses the socially exchanged information. Global best (gbest) or local best (lbest) PSO are used for finding particle's best position in the swarm. In gbest, each particle in the entire swarm acts as a neighborhood for the particle and follows the star topology of the particle's social network, where each particle is communicated with all other particles at a time. In lbest, each particle communicates with a smaller level of the neighborhood in the swarm. It follows the ring topology of the particle's social network, where it communicates with the adjacent neighbors in the swarm. The entire flow structure PSO algorithm is depicted in Figure 74. This basic PSO algorithmic step is applied over the images to identify an image's edges (Duan & Qiao, 2014).

PSO's basic and initial parameters are problem dimension, swarm size, neighborhood size, acceleration coefficients, number of iterations, and the random values used for scaling the cognitive and social components. Swarm size specifies the number of particles required for the problem. With a larger swarm size, larger search space can be covered for deeper exploration of the problem. Neighborhood size determines the societal communication level of particles inside its swarm community. As the neighborhood size is smaller, communication occurs with smaller neighbor group and has slower convergence and a lesser amount of susceptibility towards local minima. Depending upon the nature of the problem, a number of iterations can be determined. Lesser number of iterations may prematurely stop the search for the optimal solution. In vice versa, the larger number of iterations might increase the computational complexity unnecessarily (Mirjalili et al. 2014).

Other important parameters that are needed to be initialized are two acceleration coefficients and two random vectors. These factors are used for stabilizing and scaling. The initial current velocity of the particle is updated using the following equation,

$$v_i(M+1) \leftarrow wv_i(M) + x1.A(M).\left(P^{(i)}(M) - s_i(M)\right)$$
$$+ x2.B(M).\left(G - s_i(M)\right)$$
$$v_i(M+1) \leftarrow wv_i(M) + x1.A(M).\left(P^{(i)}(M) - s_i(M)\right)$$
$$+ x2.B(M).\left(G - s_i(M)\right)$$

where $v_i(M+1)$ is the i_{th} particles' change in velocity at $M+1$ iteration and $v_i(M)$ is the velocity of the particle at M_{th} iteration, w is a weighting function with $x1$ and $x2$ as an accelerating factor of G (Global best value in the swarm) and P (particles so far best value), A and B are the random numbers between 0 and 1 used as a scaling factor for cognitive and social component values of the particles' velocity and s_i denotes the position of the particle.

The position of the particle is updated using the following equation,

$$s[M+1] = s[M] + v[M+1]$$

where $s[M+1]$ is the particles' position at $M+1$ th iteration and $s[M]$ is the particle position at M_{th} iteration and $v[M+1]$ is the velocity of a particle at $k+1$ th iteration. These velocity and particle positions are updated through iterations until an optimal solution is reached.

7.10.4. Flower Pollination Algorithm

The flower pollination algorithm outperformed the standard nature-inspired metaheuristic algorithms such as the simulated annealing algorithm, genetic algorithm, firefly algorithm and bat algorithm due to its fewer parameterization and lesser time and space complexity. It is primarily reported to solve problems like nonlinear design benchmarks, disc brake design problems (multi-objective optimization problems in engineering), continuous function optimization problems and various structural engineering problems. Due to its successful and efficient performance in varied applications, the flower pollination algorithm is applied to edge detection algorithms (Yang, 2012).

The steps involved in developing an edge detection algorithm using the flower pollination algorithm are depicted diagrammatically in Figure 75.

The first and foremost task in the proposed algorithm is to initialize parameters by considering the following four main rules of flower pollination, as detailed below.

1. *In the global pollination process (Biotic and cross-pollination), levy flight is calculated based on the movement of pollen-carrying pollinators to a long distance.*
2. *In the local pollination process (abiotic and self-pollination), local random walk is calculated as it mimics the characteristics of flower constancy limited to the local neighborhood area.*
3. *The reproductive probability of two flowers determines flower constancy.*
4. *Switch probability is used to determine and control the type of pollination, either local or global.*

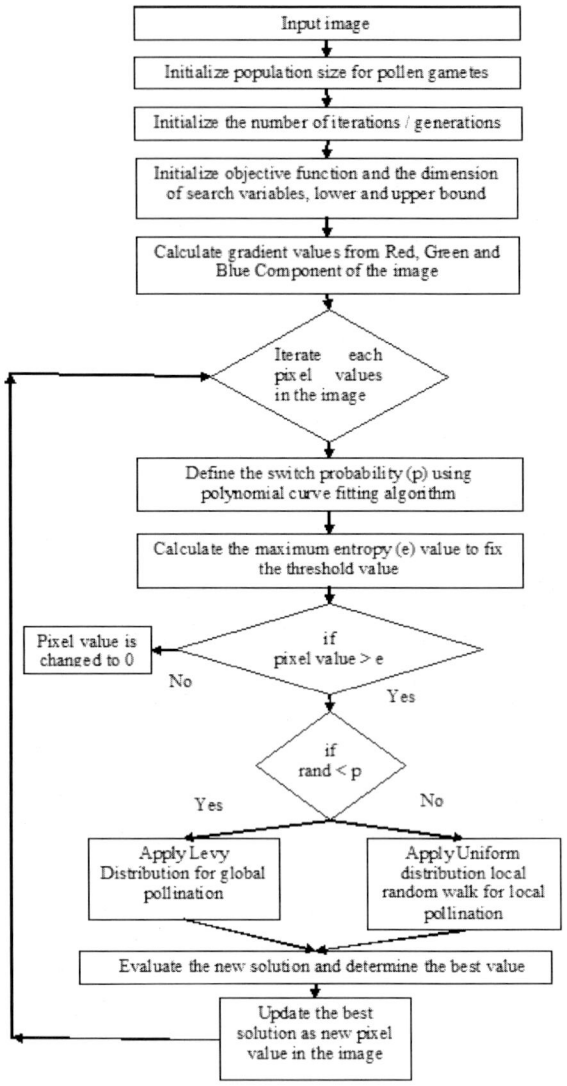

Figure 75. Steps involved in the proposed methodology.

Initialization of the population size based on pollen gametes, the number of iterations, and the dimension of search variables with lower and upper bound values are the initial task in the algorithm. The best solution from the initial population is identified using the fitness function. There are numerous in-built functions available to calculate the fitness value, but the important thing that is to be focused on is the capability of the algorithm to face higher-dimensional problems. Therefore in this algorithm rosenbrock's function is used to calculate the fitness function as it suits the dimensionality of the search and is calculated using the formula,

$$f(x) = \sum_{i=1}^{n-1}[(a_i - 1)^2 + 100(a_{i+1} - a_i^2)^2]$$

where 'n' is the number of variables, '-5 ≤ a_i ≤ 10 (i = 1,2,....n) is the search domain and the global minimum occurs at the point x* = 0.

Then the color gradient value for the input image is calculated to identify the location of images with rapid change in intensity value. It is calculated from the red, green and blue color components of the image based on the horizontal and vertical directions, respectively,

$$M = \frac{\partial I}{\partial x}i + \frac{\partial J}{\partial x}j + \frac{\partial K}{\partial x}k$$
$$N = \frac{\partial I}{\partial y}i + \frac{\partial J}{\partial y}j + \frac{\partial K}{\partial y}k$$

where I(x,y), J(x,y) and K(x,y) are the three color components of the input image E (x,y) and i,j and k are considered as the unit vectors for the three components

Then the dot product for these vectors is calculated using

$$A = M^T M = \left|\frac{\partial I}{\partial x}\right|^2 + \left|\frac{\partial J}{\partial x}\right|^2 + \left|\frac{\partial K}{\partial x}\right|^2$$
$$B = N^T N = \left|\frac{\partial I}{\partial y}\right|^2 + \left|\frac{\partial J}{\partial y}\right|^2 + \left|\frac{\partial K}{\partial y}\right|^2$$
$$C = M^T N = \frac{\partial I}{\partial x}\frac{\partial I}{\partial y} + \frac{\partial J}{\partial x}\frac{\partial J}{\partial y} + \frac{\partial K}{\partial x}\frac{\partial K}{\partial y}$$

The maximum rate of change of the function E(x,y) is given by

$$\theta(x,y) = \frac{1}{2}\tan^{-1}\left[\frac{2C}{A-B}\right]$$

The rate of change of E (x,y) in the direction of θ (x,y) is calculated using

$$D_\theta(x,y) = \{\frac{1}{2}[(A+B) + (A-B)\cos 2\theta\,(x,y) + 2C\sin 2\theta\,(x,y)]\}^{\frac{1}{2}}$$

and finally, magnitude is calculated using

$$Mag = \sqrt{(D_\theta)}$$

Thus the color gradient image is derived for the input RGB image.

The subsequent important step is to calculate the maximum entropy for the gradient image. Entropy, in general, is used to measure the randomness value in an image. The principle of a maximum entropy probability distribution is to measure the low and high range of entropy values and the maximum value is used to determine the threshold value. This information of entropy is calculated using the formula,

$$I = \sum_i p(S_i) \log_2\left(\frac{\sum_i p(S_i)}{p(S_i)}\right)$$

where S_i is the possible states and $p(S_i)$ is the probability of possible states.

The next step is to start the loop process for the entire image pixels to identify the exact edge region in an image. Based on the switching probability defined, the best solution calculation is switched between levy flight distribution and local random walk distribution. Switch probability is defined for the problem based on the polynomial curve fitting distribution. The intensity distribution of the image is normalized to the value of 0 and 1. The normalized intensity value closer to the curve is identified and is considered for defining the switching probability. If the random value is less than the switching probability defined, then the levy flight distribution equation is used to calculate the new solution based on global pollination and the equation is as follows,

$$x_n^{i+1} = x_n^i + L(x_n^i - b)$$

where the subscript 'n' represents the pollen and the subscript 'i' represents the iteration step and x_n^i is the solution vector for n^{th} pollen at i^{th} iteration and Levy distribution is calculated using the formula

$$Levy \sim \frac{\lambda G(\lambda) \sin\left(\frac{\pi\lambda}{2}\right)}{\pi} \frac{1}{G^{1+\lambda}}, (G \gg G_0 > 0)$$

where $G(\lambda)$ is the gamma function and the gamma value assigned by default is 1.5. This distribution is valid for values of $G > 0$.

If the value is greater than switch probability then local random walk uniform distribution is used to calculate the new solution based on local pollination and the equation is as follows,

$$x_n^{i+1} = x_n^i + \varepsilon(x_m^i - x_t^i)$$

where x_n^i and x_n^i are the pollens arriving from different flowers but the same plant species.

The new solution is now evaluated with the objective function to check if it is better than the existing one. If it is better, the new value is updated as the global best and is updated as the new best edge pixel value in an image else, the pixel value is considered zero. This process is iterated for all the pixel values in an image to obtain a better edge detected image.

Chapter 8

Feature Extraction Techniques Applicable for Agricultural Images

A set of information extracted from the given source image is referred to as features. Certain properties are required for a feature. It must be reliable, robust, compact and discriminant. Three basic types of features can be obtained from an image, namely low-level, middle-level and high-level features. The basic information extracted like color value and gradient value from an image are considered low-level features. If the information are extracted through edges, regions and corners, then this type of information is considered as a mid-level features. The deformable active contour model is referred to as high-level information. It is sufficient to extract the image's low and middle level features. Shape, spatial and texture are the major descriptors of these features. The basic shape features are area, perimeter, length, width and dimensionless shape features are roundness, elongation and compactness. The basic spatial features are the intensity value information obtained through statistical information˙ like mean, standard deviation and variance. The basic texture features are entropy, correlation, homogeneity and contrast. To extract the exact information from both agriculturally important images, the low and middle-level features are used. Properties like color, shape and texture are obtained from the image.

8.1. Color Value Extraction

The primary feature for the classification of the maturity level of banana was the color. The captured image was in an RGB color model. Hence the RGB intensity color distribution in the banana was calculated using statistical moments obtained from the histogram. The histogram for the banana region was computed using the vertices stored in the vectors (Quevedo et al. 2008). The horizontal axis of the histogram represents the intensity scale ranging from 0 to 255. The vertical scale of the histogram represents the frequencies of the pixel distribution. The histogram measures the pixel distribution on the intensity scale of 0 to 255. The central moment $\mu`$ has been calculated, which provides quantitative information about the shape of the histogram as pixel distribution. Central moments also referred to as moments about the mean, have been calculated as,

$$\mu_m = \sum_{n=0}^{L-1} ((x_n - y)^m t(x_n))$$

where 'm' is an order of the moment, 'L' is the number of possible intensity values, 'x_n' is the discrete variable representing an image's intensity level and 'y' is the mean. $t(x_n)$ is the probability estimation of the occurrence of. 'x_n'.
Mean can be defined as,

$$y = \sum_{n=0}^{L-1} (x_n \, t(x_n))$$

From equations (1) and (2), the moment for order '0' was always '1' and the moment for order '1' was always '0'. Hence these moments were ignored. The mean was considered first-order moment, followed by variance, skewness and kurtosis as the second, third and fourth moment. Mean in the first order central moment was used to measure the average intensity value of the pixel distribution. The mean helps to identify the average color value of the pixels in an image.

Variance (μ_2) was used to measure how wide the pixels spread over from the mean value.

$$\mu_2 = \sum_{n=0}^{L-1} ((x_n - y)^2 t(x_n))$$

The smooth texture of an image was measured using a variance. The image with constant intensity had a smoothness of '0' and the image with varying intensity had a smoothness of '1'. In general, the smoothness of an image lies within '0' to '1'. Smoothness texture 'R' can be defined as,

$$R = (1 - (1/(1 + \mu_2(x))))$$

where 'μ_2' is the variance and 'x' is an intensity level.

Skewness(μ_3) was used to measure the asymmetric shape of the distribution. This shape distribution was either positively or negatively skewed. The distribution positively skewed denotes the left side of the mean had more pixel values clustered, but the tail extends towards the right side of the mean as in Figure 76(a). Similarly, distribution negatively skewed denotes right side of the mean had more pixel values clustered, but the tail extends towards the left side of the mean, as in Figure 76(b).

$$\mu_3 = \sum_{n=0}^{L-1} \left((x_n - y)^3 \, t(x_n) \right)$$

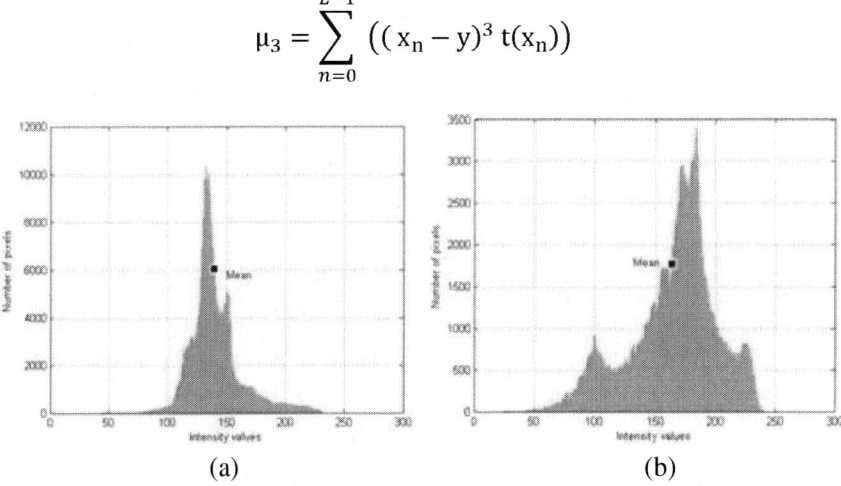

(a)   (b)

Figure 76. (a) Positive skewness and (b) negative skewness distribution.

Kurtosis (μ_4) was used to measure the peakedness of the distribution. The distribution of the shape was either positive or negative. Kurtosis with a positive value was referred to as leptokurtic. It had higher peaks around the mean with heavy tails, as in Figure 77(a). Kurtosis with a negative value was referred to as platykurtic. It had flatter peaks around the mean with thinner tails, as in Figure 77(b).

$$\mu_4 = \sum_{n=0}^{L-1} \left((x_n - y)^4 \, t(x_n) \right)$$

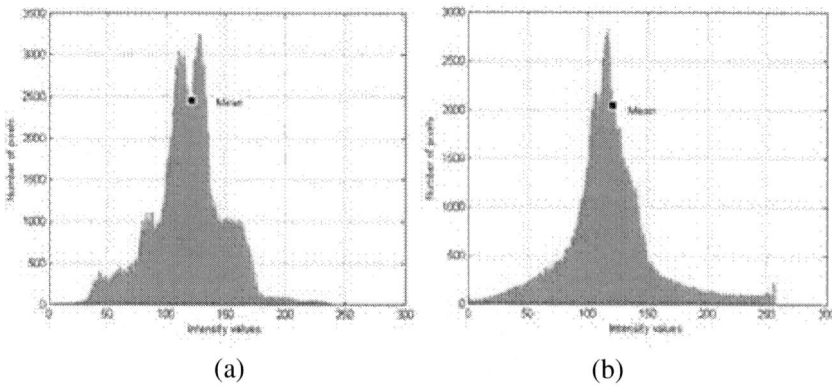

Figure 77. (a) Positive kurtosis and (b) negative kurtosis distribution.

8.2. Size Value Extraction

The size of fruit or vegetable is the second important feature in determining maturity as well as quality. The area and perimeter of the agro product are generally used to represent their size feature. The area of the produce is measured using a total number of pixels in the target region. The target's perimeter can be measured using the total number of pixels in the boundary region of the product. Major axis length and minor axis length of the agro produce is also measured using regional descriptors available in an image processing toolbox of Matlab. The measurement units of area, perimeter, major axis length and minor axis length of the agro-produce image are in pixels. These pixels will be converted into a centimeter measurement unit for easy recognition by farmers on the target size. Conversion is made by using a reference object with known size values for a standard pixel count. The cross-sectional cut of the agro produce can be used as the reference object. The blueprint of the cross-sectional part is taken in a graph sheet. The area and perimeter of the cross sectional parts are measured using the coordinate geometry from the graph sheet. The area of the cross-sectional produce using the co-ordinate chain values can be calculated as;

$$\begin{aligned}Area\ of\ polygon &= 1/2[(x_0 y_1 - x_1 y_0) + (x_1 y_2 - x_1 y_0) \\ &+ (x_2 y_3 - x_3 y_2)(x_3 y_0 - x_0 y_3)\end{aligned}$$

Area of polygon
$$= 1/2[(x_0y_1 - x_1y_0) + (x_1y_2 - x_1y_0) + (x_2y_3 - x_3y_2)(x_3y_0 - x_0y_3)]$$

where (x0,y0), (x1,y1), (x2,y2), (x3,y3) are the co-ordinate values of the polygon.

The length and width of the reference banana object were measured using the Pythagoras theorem (Vince 2010). The distance between two points in the coordinate geometry is represented as;

$$Distance = \sqrt{\Delta x^2 + \Delta y^2}$$

where $\Delta x = (x_2 - x_1)$ and $\Delta y = (y_2 - y_1)$.

The intersection points of (x1, y1), (x2, y2) were considered for the distance measurement.

An object's perimeter was calculated as the sum of the object's length from each side.

The image of the reference object is taken and its values in the pixel are calculated for the area, perimeter, major axis length and minor axis width. The measurement unit in centimeter and the pixel value of the reference object are compared and the value of one pixel in centimeter is measured and used to calculate the size of the agro-produce in centimeters (Vince & Vince, 2010).

8.3. FAST Algorithm

A FAST algorithm is used to identify the exact and definite corner or interest points in an image (Prabha, 2017). These interest points are important in an image as it provides detailed local information about the pixel points and this algorithm is explained below:

a. A pixel point 'p', is selected in the image with an intensity value of I_p to determine if the pixel is an interest point by setting up a threshold value for selection.
b. Using the Bresenham circle algorithm, a circle comprising 16 pixels around the point p is considered for manipulation.

c. Among the 16 pixels, N adjoining pixels are selected to determine whether the point p is an interest point. For this purpose, it checks either of two conditions. In condition1, if the N set of adjoining pixel intensity value along with the threshold value is brighter than the pixel point (p) then the point (p) is considered as a corner point or interest point. In condition2 N, if the adjoining pixel intensity value set along with the threshold value is darker than the pixel point (p). The point (p) is considered a corner point or interest point. If both conditions are not satisfied, then the pixel point p is not considered an interest point or corner point.
d. This procedure is repeated for all the pixels in an image to determine the interest points for the entire image.

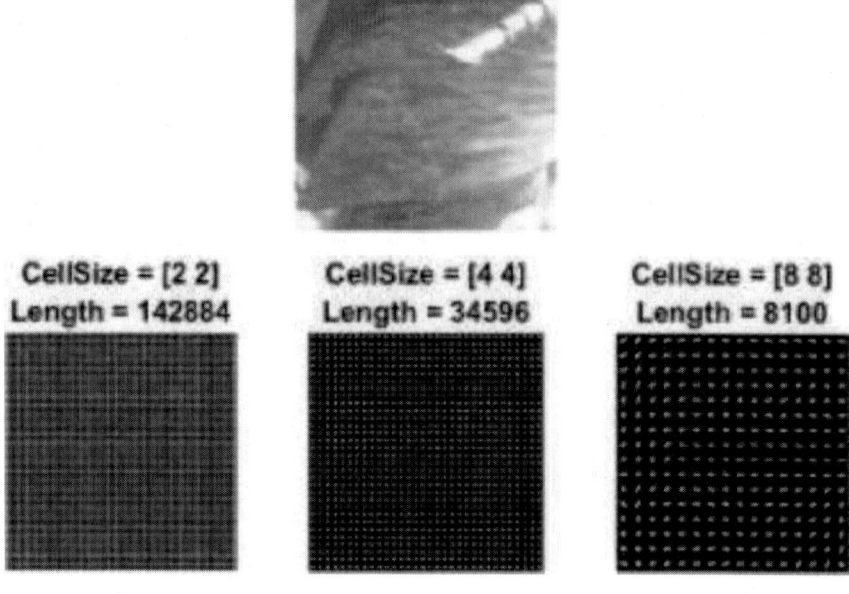

Figure 78. Distribution pattern of HoG gradient features.

8.4. Histogram of Gradients

After extracting the interest points in an image, the next step is to determine the feature descriptors to make the classification process easier and more efficient. HOG is commonly referred to as a dense feature extraction method due to its characteristics of extracting dense features from the entire locations in an image. The basic idea behind this algorithm is to extract the shape structure of an image from its gradient value. The steps involved in this algorithm are described below:

a. The image is divided into a number of small regions called cells and for each of these cells horizontal and vertical gradient of the image is calculated using the formula,
b. Then normalization and orientation of the gradient are calculated using the formula,
c. Next, the histogram is calculated for each cell and is discretized based on histogram angular bin values.
d. In the next step, adjacent cells are grouped as blocks and form the base for grouping histograms and normalization.
e. Finally, all the histograms within cell blocks are grouped and normalized to generate a single vector as the final descriptor, as in Figure 78.

8.5. Gray-Level Co-Occurrence Matrix (GLCM)

The gray-level co-occurrence matrix (GLCM) is a feature extraction techniques useful in agricultural images like root discoloration due to pathogens and nematodes. Size, shape, texture and color are the important features that are needed to be extracted from the root images. In root images, the color varies from healthy roots to slightly infected roots and dead roots, as in Figure 79. The color of healthy roots is usually in white color space, the infected root range in pale yellow, white with dark black marks, light brownish shades and the dead roots are usually in black and dark brown color. So color features play a vital role in unique root classification. The size and shape feature here does not impact the classification of root diagnosis (Suryaprabha et al. 2020).

The texture is another important feature that plays a vital role in root disease diagnosis. Gray-level co-occurrence matrix (GLCM) is a statistical method used to extract texture features based on the spatial relationship among the pixels. Contrast, correlation, energy and homogeneity are some statistical features that provide information about texture. These features are extracted with the GLCM function. The contrast feature provides information related to local variations among the pixels in an image. Correlation provides information related to joint probability occurrence among the pixels in an image. Energy provides information about the uniformity among the pixels in an image. Homogeneity Mean, standard deviation and variance are some of the statistical features that help identify the mean's central moment for an image through its histogram distribution.

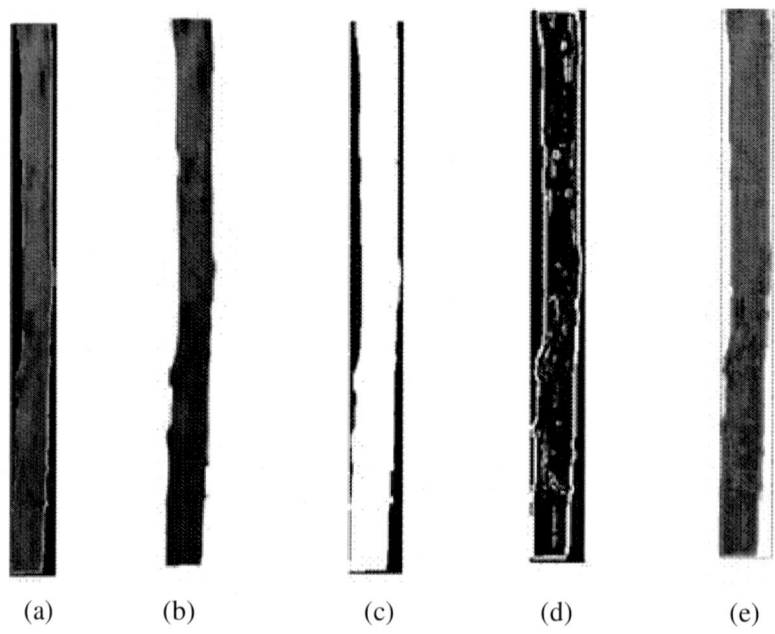

Figure 79. Sample image of diseases identified from banana root disease dataset a) infected root, b) background removed root, c) background part, d) gray level co-occurrence matrix and e) color feature extraction after gray level co-occurrence matrix

Chapter 9

Image Classification or Pattern Recognition to Solve Agricultural Problems

Image classification is a higher level of image processing module which deals with the concept of machine learning algorithms and plays a vital role in object detection and recognition. So this module plays a vital role in identifying plant, weeds, pest and disease monitoring and solving many problems in agriculture. Machine learning algorithms are computer program that learns from the training dataset and their performance is improved based on the experience it acquires through training. Supervised, unsupervised and semi-supervised are the major categories of machine learning algorithms (Figure 80) (Mathews & Fink, 2004).

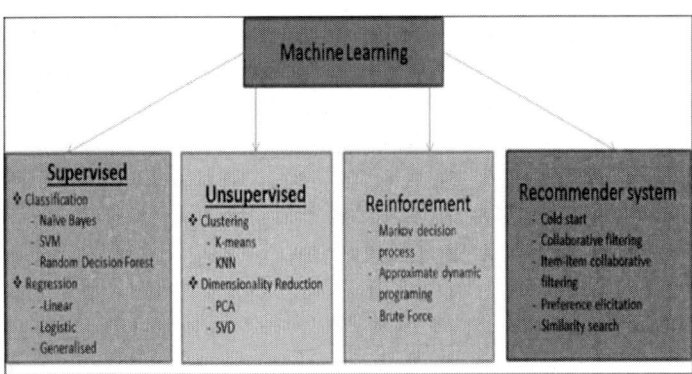

Figure 80. Kinds of machine learning techniques.

If the datasets classes are of known label classes, supervised learning algorithms are considered suitable. Classification algorithms learn from the labeled training observations that are used in the training datasets. A many number of different known labeled classes can be used in the dataset. Popularly Support Vector Machines, Decision trees and Convolution network algorithms are now being used for solving problems. Before implementing the classifier algorithm, the dataset of each category is divided into two sets: training set and testing or validation set. The 70% of images of each category are used as training data and for testing the developed classifier model, the remaining 30% of data are used.

9.1. Support Vector Machine

SVM is a supervised discriminative classification algorithm designed for binary classification. However, its robustness in solving varied problems has made the researchers expand it for multi-class classification. During its implementation for multi-class classification, the problems need to be decomposed into a number of binary problems. Error Correcting Output Codes (ECOC) method decomposes the multi-class into a set of binary classes for classification. Two types of ECOC methods, 'one-versus-one' and 'one-versus-all' for decomposing the multi-class classification. One-versus-one approach can be used for multi-class SVM in which classification was done based on the max-wins vote strategy (Prabha, 2017).

In the SVM algorithm for binary classification, the initial step is finding an optimal hyperplane, separating the binary classes using the maximum margin rule. This optimization problem of finding the maximized margin criteria for a training vector ($y_i \in T^d$, i = 1,...n, in two classes) with label vector ($L \in \{1, -1\}^n$) is done by solving this mathematical equation,

$$\min_{v \in F, a \in R, \varepsilon_i \in R} \frac{1}{2} v^T v + C \sum_{i=1}^{l} \xi_i \tag{1}$$

Subject to

$$L_i(v^T \varphi(y_i) + a) \geq 1 - \xi_i \tag{2}$$

where $v \in T^d$ is weight vector, C is a regularization constant, φ is a mapping function used to map or convert the training data into a feature vector space to support nonlinear decision surfaces.

In pairwise decomposition or one-versus-one multi-class classification, it uses k (k-1)/2 binary classifiers individually and evaluates all these possible pairwise classifiers. Each classifiers is applied to test samples and voting is done for the winning class. These created classifiers are much larger and symmetric than the one-versus-rest classifier approach.

9.2. Principal Component Analysis

Before implementing PCA, an instance of a model is generated first and all parameters not specified are set to their defaults. Then standardization of data is done secondly. The features of the image are scaled and standardized before applying PCA. Standard Scaler is used to standardize the dataset's features onto the unit scale (mean = 0 and variance = 1) to meet the requirement for the optimal performance of machine learning algorithms. Then, a popular unsupervised machine-learning algorithm PCA can be performed. Hence, the original image data of three dimensions or components, are converted into two dimensions. After dimensionality reduction, the new components are just the two main dimensions of variation. During the implementation of PCA, repeated and unwanted information is removed. Only useful information is collected from the data set.

Chapter 10

Quality Assessment Metrics of Reference and Non-Reference Agriculturally Important Images

The quality of the output image obtained through various image-processing steps are needed to be assessed quantitatively. Generally, evaluation methods are classified into two broad categories: subjective and objective, as in Figure 81 (Avcibas et al. 2002).

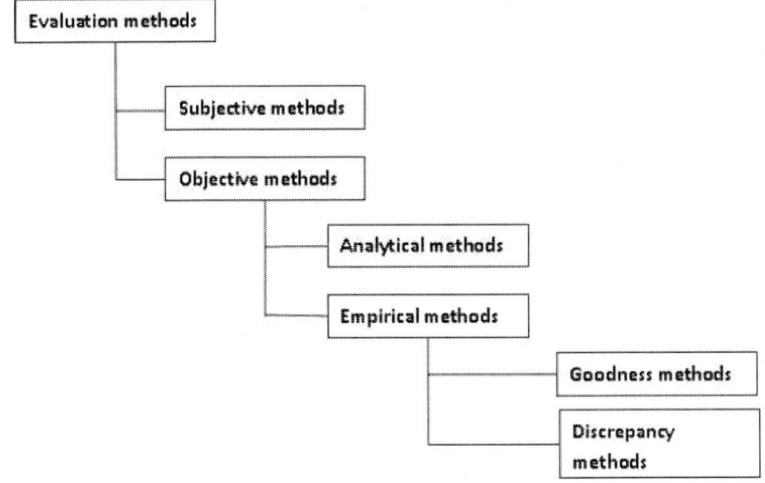

Figure 81. Different image evaluation techniques.

It is a usual practice to evaluate the quality of output images obtained after imposing various image processing steps by the subjective assessment method. The subjective method is an evaluation based on human visual inspection, which is biased or unreliable, time-consuming and expensive. So it has become essential to have a reliable, accurate and unbiased objective assessment method. The objective method does not involve human assumptions and assessments. It is further classified into two groups analytical method and empirical method. Most of the image processing methods are evaluated based on the empirical, indirect method of evaluation. The goodness evaluation method and discrepancy evaluation method are the two major classifications of empirical method based on the use of reference, ground truth, or gold standard image. Discrepancy evaluation methods, also known as supervised or relative evaluation methods, evaluate the performance of image processing algorithms by analyzing the similarity between enhancements or segmentation algorithms applied output image and the ground truth image (Cardoso & Corte-Real, 2005). Human experts usually generate ground truth images. However, it is possible to generate the ground truth image only for synthetic images.

Therefore, for real time images, it is impossible to compare the image processing output developed for specific applications with ground truth images. In these cases, the goodness method, the stand-alone or unsupervised evaluation method, is used to evaluate the performance accuracy of image processing like segmentation algorithms (Zhang et al. 2008).

10.1. Subjective Evaluation Method

Subjective evaluation is a commonly used performance assessment method in the literature. Human involvement is the basic requirement in this method as assessment is determined based on human visual inspection. A major challenge in this method is the varying result of human inspectors. Evaluation result obtained from this method is biased, expensive, time-consuming and the probability of accuracy is low in this method. Another major drawback is the requirement for a large number of human inspectors. Parameter selection is another problem faced in this method as it is biased and based on favoritism.

10.2. Objective Evaluation Method

This method provides a reliable comparison among the enhancement or segmentation algorithms. It compares the performance of enhancement or segmentation methods with the golden standard based on properties of the image like distance and similarity measure. It imitates certain characteristics from subjective methods and uses human expertise. Analytical and empirical methods are the two main classifications in objective evaluation.

10.2.1. Analytical

Analytical evaluation methods require prior knowledge to evaluate the enhancement or segmentation algorithms by considering their nature, needs, and complications characteristics. This method is complex and complicated to analyze and compare the algorithm's performance, as it is not reliable and consistent. Due to a lack of proper theoretical knowledge of enhancement or segmentation and the inability to extract all features from an image, this method is not preferred for evaluating the performance of enhancement or segmentation algorithms (Heath, 1998).

10.2.2. Empirical

Generally, empirical evaluation methods are the widely used evaluation technique to measure the performance of enhancement or segmentation algorithms. This technique uses an enhancement or segmentation algorithm on test images to evaluate the performance. This method is simple, faster, and reliable to produce accurate results. This method is competent to evaluate numerous sets of out images automatically in a smaller period. The empirical method evaluates the images based on goodness or discrepancy measures (Ji & Haralick, 2002).

10.2.2.1. Discrepancy Measures

The discrepancy evaluation method, also called relative or supervised evaluation method, is used to evaluate the performance of enhancement or segmentation methods based on the concept of using reference or ground truth image. In order to define a ground truth image, human expertise is needed to have a hand-drawn output result. This method is suitable in cases of images where images are predetermined and their golden standard images are generated with the knowledge of human expertise (Yitzhaky & Peli, 2003). This method measures the relationship between the output image of enhancement or segmentation method and the ground truth image. Discrepancy methods are broadly categorized into three groups: similarity, distance, and standard, as in Figure 82. This evaluation method is considered to produce an evaluation result with higher accuracy. One of the shortcomings of this method is the generation of ground truth, which is time-consuming, biased and tricky. Some of the commonly used discrepancy methods are discussed in detail.

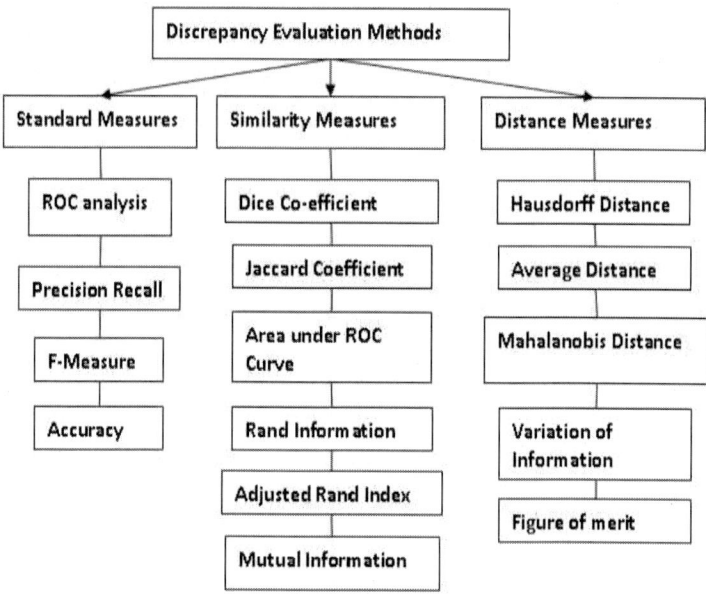

Figure 82. Different discrepancy evaluation techniques.

Receiver Operating Characteristics (ROC) Curve
ROC curve is a pixel-based standard measure used to compare the ground truth image and output image of the segmentation method based on a confusion matrix. Factors involved in the confusion matrix generation are true positives (TP), false positives (FP), true negatives (TN) and false negatives (FN), as in Figure 83. Sensitivity and 1-specificity are two measures required for plotting the ROC curve.

Sensitivity or true positive rate or recall is the percentage of true positive pixels and its formula is,

$$Recall = TP/((TP + FN))$$

1-Specificity or fallouts or False Positive rate or fallout is the percentage of false positive pixels and its formula is,

$$Recall = (TP)/(TP + FN)$$

A higher percentage of sensitivity and 1-specificity assures that the segmentation method is of good quality and has higher perfection.

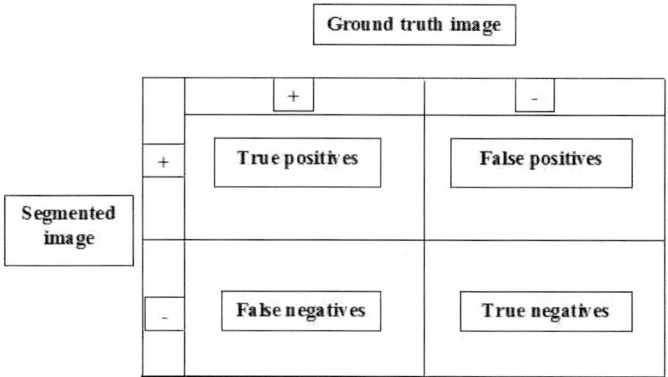

Figure 83. A sample confusion matrix.

Area under ROC Curve (AUC)

It is a simple measurement metric used to measure the accuracy by reducing the ROC curve result into a scalar value. The value of this method is normalized between the range of 0 and 1. A higher value of AUC indicates a better performance of the segmentation. It is calculated using the formula,

$$AUC = \int_x^y f(a)da$$

where 'x' and 'y' are the curve's minimum and maximum axis points with 'f(a)' a function partly above and below the curve. In simple words, AUC is the difference between the area above the ROC curve and the area below ROC curve.

Precision-Recall (PR) Curve

Precision-recall curve is a standard evaluation technique similar to Receiver Operating Characteristics (ROC) curves. It is also a pixel-based measure used to evaluate the algorithm's performance. It uses the confusion matrix to match the relationship between the ground truth image and segmented output image based on the edge pixels and non-edges in these images. This technique is popularly used in information retrieval and pattern recognition. Precision or positive predictive value) is a percentage of true positive pixels that are relevant and recall is a percentage of true positive pixels that are retrieved. As the ROC curve uses fallout and recall for plotting the curve, the precision-recall curve uses the positive predictive value and recall for plotting the curves. A positive predictive value is calculated based on the formula

$$PPV = \frac{True\ Positive}{True\ Positive + False\ Positive}$$

The concept of precision-recall has been introduced for segmentation evaluation as it is considered a positive predictive value for analyzing instead of a false positive rate which is not appropriate for evaluating the quality of the segmentation algorithm. Precision gives information about the validity of segmentation results and recall provides information about the correctly identified edge pixels in an image. A higher value of precision and recall indicates a good performance by the segmentation method. Under segmentation results in the segmentation method when the value of recall is low and over segmentation results when the value of precision is low.

F-Measure
F-measure is used to measure segmentation efficiency and success based on precision and recall values. To have a single measure with higher effectiveness, a unimodal F measure is calculated by combining precision and recall. It is a harmonic mean that gives a precise result and is defined using the formula

$$F\ Measure = 2.\ (precision\ .\ recall)/(precision + recall)$$

In the precision recall curve, the value of the F-measure can be pointed and in most situations, F-measures are marked in the PR curve related to the segmentation methods. Its value ranges from 0 to 1. The segmentation method with better performance has a higher value when compared and evaluated with other segmentation methods (Zhang, 1996).

Accuracy
The accuracy of the segmentation method is defined as the ratio between the number of correctly matched pixels and the total number of pixels. It is determined by using the formula

$$Accuracy = (TP + TN)/(TP + FP + TN + FN)$$

Figure of Merit
It is another practical measure used for evaluating the edge-based segmentation method's performance. It is based on the mean-square distance between all pixel pair points in the segmented output image and ground truth

image and assesses their similarity. This method of evaluation proposed by Pratt is not only useful for assessing the quality of edges but is also useful for assessing the entire behavior of the segmentation method. Its value ranges from 0 to 1, representing optimal segmentation result. It is calculated and evaluated using

$$Figure\ of\ merit = (\frac{1}{\max\{N_g N_d\}} \sum_{i=1}^{N} \frac{1}{1 + C * D_i^2})$$

where 'N_g' is the number of edge pixels in the ground truth image and 'N_d' is the number of edge pixels in the segmented output image. D is the distance between the detected edge pixel point and its accurate edge pixel point (Zhang, 1997).

Area under ROC Curve (AUC)
It is a simple measurement metric used to compare the performance of segmentation methods and to measure accuracy. This method is used to ease ROC performance by reducing the result into a scalar value. The value of this method is also normalized between 0 and 1. A higher value of AUC indicates a better performance of the segmentation. It is calculated using the formula,

$$AUC = \int_{x}^{y} f(a)da$$

where 'x' and 'y' are the curve's minimum and maximum axis points with 'f(a)', a function partly above and below the curve. In simple words, AUC is the difference between the area above the ROC curve and the area below ROC curve.

Rand Index
It is a similarity measure used to analyze the similarity between the ground truth and output image of the segmentation method. It is a common method used in data clustering to measure the similarity among different clusters. It is a method that is associated with accuracy. In segmentation evaluation, it computes the percentage of reliability and consistency between the two pairs of images. It is calculated using the formula

$$Rand\ Index = \frac{(m+n)}{(m+n)+(o+p)}$$

where 'm' is the number of pair of pixels that are the same between two images, 'n' is the number of pair of pixels that are different between two images, 'o' is the number of pair of pixels found in ground truth image and not in the segmented image, 'p' is the number of a number of pair of pixels not found in ground truth image and number of pair of pixels in the segmented image, (m+n) is the number of correct agreement between ground truth image and segmented image and (o+p) is the number of wrong agreement between ground truth image and segmented image (Zhu et al.1999).

Variation of Information (VI)
Variation of Information is a technique commonly used in comparing the clustering method. It is used to measure the distance between two images, i.e., the ground truth image and segmented image and thereby, it calculates the randomness of information in one image that is not available in another image. This method calculates the randomness of information using the conditional entropy and the formula for VI is defined as follows

$$VI = E(X) + E(Y) - 2M(X,Y)$$

and mutual information M(x,y) is calculated using

$$M(X,Y) = \{E(X), E(Y)\} - \log_2 n$$

where 'X' and 'Y' are the ground truth image and segmented image, E(X) is the entropy value of X, and E(Y) is the entropy value of Y and M (X,Y) is the mutual information of X and Y and 'n' is the total number of pixels in the image.

Dice-Coefficient
Dice coefficient is a similarity measure that calculates the proportion or percentage of similarity between the ground truth image and segmented image. This method is mostly used in medical image processing to evaluate the performance of segmentation algorithms with a predefined ground truth information or data set. It is calculated using the formula

$$\text{Dice co-efficient} = \frac{2\,|X \cap Y|}{|X| + |Y|}$$

where X was the non-zero pixel element in the ground truth image and Y was the non-zero pixel element in the segmented image.

Jaccard-Coefficient
Jaccard Co-efficient is also similar to that of the Dice coefficient used to calculate the similarity between the two set of images and it also measures the variation or dissimilarity between two image. It is calculated using the formula

$$\text{Jaccard co-efficient} = \frac{|X \cap Y|}{|X \cup Y|}$$

and Jaccard distance is calculated using

$$\text{Jaccard distance} = \frac{|X \cup Y| - |X \cap Y|}{|X \cup Y|}$$

where X was the non-zero pixel element in the ground truth image and Y was the non-zero pixel element is the segmented image.

10.2.2.2. Goodness Measures
The goodness method, also known as unsupervised or stand-alone method of evaluation, is used to evaluate the performance of segmentation based on the segmentation image characteristics. This method of assessment does not require any predefined ground truth image and prior knowledge for evaluation. This method is very useful for situations where it is impossible to collect the ground truth images. Properties of images like shape, region, color, texture, variance, uniformity and entropy are used as key factors to analyze the performance of segmentation methods. Based on these properties, goodness measures are classified into different methods, as in Figure 84. Some of the frequently used goodness measures are discussed in this section.

Cohens' Kappa
It is a statistical method used to evaluate judgment based on the resultof different persons by analyzing the level of agreement among those persons. This method is meaningful as it considers not only the observed agreements but also the probability of agreements by chance. In image segmentation, it is

a pixel-by-pixel comparison. It considers and compares the pixels in segmented region of an image and the probability of pixels found in the segmented region of an image. Kappa value is calculated using the formula,

$$Kappa = (O - E)/(1 - E)$$

where O is the observed pixels in the segmented image and E is the probability of having a pixel by chance. The value of kappa is normalized to range from 0 to 1. Higher the value of kappa, the better the performance of the segmentation method.

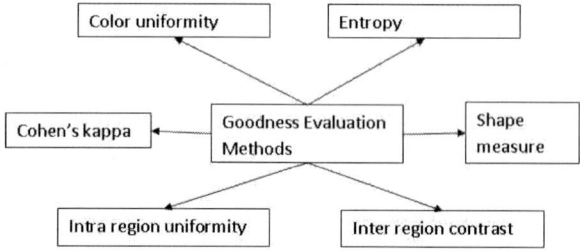

Figure 84. Different goodness measure methods.

Entropy
Entropy is another evaluation measure used to compute an image's randomness or information content. It assists in calculating the uniformity measure in an image.
  Shannon entropy was calculated using

$$Entropy = -\sum_{i=1}^{a} e_i \log e_i$$

where 'e' represents the pixels frequency and 'I' represents the pixel's intensity value. The lower value of entropy assures less randomness in the image information and vice versa for a higher value of entropy which shows more randomness in the image. Therefore, for the segmentation method with better performance, the entropy value will be lesser and the entropy value will be higher for the poor performance method.

Shape Measure
It is used to evaluate the performance of the segmentation method based on shape features. It uses the gradient value and neighborhood pixel to determine the accuracy. It is calculated using the formula as follows

$$M = \frac{1}{c} \{\sum_{(a,b)} s[I(a,b) - I_{J(a,b)}] G(a,b) s[I(a,b) - x]\}$$

where c is a constant scalar value, s is an element-wise step function, $I(a,b)$ is the grayscale image, $I_{J(a,b)}$ is the neighborhoods average value at each pixel location of (a,b) for the image $I(a,b)$ and $G(a,b)$ is the gradient value for the image and x is the threshold value.

Intra Region Uniformity
It is used to analyze the characteristics of a segmented image based on its region uniformity. Inter-region uniformity and Intra region uniformity is calculated for the segmented image in both their foreground image and background image. These values are analyzed by selecting an appropriate threshold value that exactly distinguishes an image's foreground and background. Busyness is another feature used to evaluate the performance of the segmentation method as it assumes that both background and foreground objects in an image are solid in shape with robust texture. Performance vector is also used for evaluation, providing all information related to region uniformity, region contrast and texture.

10.2.3. Other

The other widely used evaluation techniques are Peak Signal to Noise Ratio (PSNR), Structural Similarity Index Measure (SSIM) and Feature Similarity Index Measure (SSIM), which are also considered for image quality assessment. They can be used to evaluate the output images after imposing image enchantment algorithms.

Peak Signal to Noise Ratio (PSNR)
It is an appealing conventional image quality assessment metric that calculates the error sensitivity between the original and enhanced image (Hore & Ziou,

2010). This method is designed based on simple mathematical calculation and unambiguous information and is calculated using the formula,

$$\text{PSNR}(K, I) = 10\log_{10}\left(\frac{M^2}{\text{MSE}(K, I)}\right)$$

where 'M' is the maximum pixel value recorded of a pixel in an image. Usually, 'M' is 255 for an 8-bit image and the mean square error (MSE) is mathematically expressed as follows,

$$\text{MSE}(K, I) = \frac{1}{ST} \sum_{i=1}^{S} \sum_{j=1}^{T} (K_{ij} - I_{ij})^2$$

where, 'K' is an original image and 'I' is an enhanced image with an image size of $S \times T$, 'i' and 'j' as their respective coordinate values. Image quality is excellent for enhanced images, which have a higher value of PSNR, and is poor for images with the lower value of PSNR.

Structural Similarity Index Measure (SSIM)
This method uses the structural information of an image for quality measurement of an image (This method was proposed by Wang et al. (2004) as an extension of the Universal Image Quality Index. Luminance, contrast and structure are important features to measure the similarity between two images. The main advantage of this method was its pixel-by-pixel quality analysis with easier; simpler and understandable mathematical knowledge and is calculated using the formula,

$$SSIM = \frac{(2\mu_a\mu_b + c_1)(2\sigma_{ab} + c_2)}{(\mu_a^2 + \mu_b^2 + c_1)(\sigma_a^2 + \sigma_b^2 + c_2)}$$

where 'a' is the original image, 'b' is the distorted image, 'μ_a' is the mean value of image 'a', 'μ_b' is the mean value of image 'b', 'σ_a' is the standard deviation of image 'a', 'σ_b' is the standard deviation of image 'b' and 'σ_{ab}' is the covariance value for images 'a' and 'b' with 'c_1' and 'c_2' the constant values used to stabilize the result.

The above-said mean, standard deviation and covariance values are calculated as follows,

$$\mu_a = \frac{1}{N}\sum_{x=1}^{N} a_x$$

$$\mu_b = \frac{1}{N}\sum_{x=1}^{N} b_x$$

$$\sigma_a^2 = \frac{1}{N-1}\sum_{x=1}^{N} (a_x - \mu_a)^2$$

$$\sigma_b^2 = \frac{1}{N-1}\sum_{x=1}^{N} (b_x - \mu_b)^2$$

$$\sigma_{ab}^2 = \frac{1}{N-1}\sum_{x=1}^{N} (a_x - \mu_a)(b_x - \mu_b)$$

where 'a_x' and 'b_x' denote the pixel intensity values of original image (a) and distorted image (b) with 'N' as the total number of pixels in an image. Similarity (SSIM) between two images lies within a range of -1 to 1, and original and distorted images are identical and have SSIM values equal to 1. In simple words, a distorted image with a high loss of structural image from the original image has a lower value of SSIM and vice versa for an image with less loss of structural information.

Feature Similarity Index Measure (FSIM)

It is a recent quantitative analysis metric used for image quality assessment. This method is based on the low-level features extracted from an image. Two main components, Phase Congruency (PHC) and Gradient magnitude (Grad)are used in FSIM to extract the local feature information from an image (Zhang et al. 2011). Phase congruency is a primary step used as a dimensionless measure for local feature identification and is considered a contrast invariant. This is followed by gradient magnitude, which supports balancing the mapping of local similarity features between two images. Finally, it is used again to attain a single score as resultant information. The phase congruency for a 2-D image at each position *i* on a scale *a* form a response vector $[k_a(i) l_a(i)]$ for each quadrature pair from an image and is defined as follows

$$\text{PHASECON}(i) = \frac{\sum_x Z_{\theta_x}(i)}{\varepsilon + \sum_N \sum_x P_{a,\theta_x}(i)}$$

where 'ε' is a constant value, 'θ_x' is defined as follows;

$$\theta_x = \frac{x\pi}{X}$$

where 'X' represents the number of orientations and 'x' is the orientation angle ranging from $\{0,1,....(X-1)\}$ and '$P_{a,\theta_x}(i)$' is the local amplitude of point 'i' on the scale 'N' with an orientation 'θ_x' and is defined as follows,

$$P_{a,\theta_x}(i) = \sqrt{k_{a,\theta_x}(i)^2 + l_{a,\theta_x}(i)^2}$$

and 'Z_{θ_x}' is the local energy level for the point 'i' with an orientation of 'θ_x' and is defined as follows,

$$Z_{\theta_x}(i) = \sqrt{M_{\theta_x}(i)^2 + N_{\theta_x}(i)^2}$$

where

$$M_{\theta_x}(i) = \sum_a k_{a,\theta_x}(i)$$
$$N_{\theta_x}(i) = \sum_a l_{a,\theta_x}(i)$$

The next step is to calculate the gradient of an image where a convolution mask is a traditional approach and its role is important in FSIM calculation. Sobel, Prewitt and Scharr are commonly used gradient operators for calculation. Gradient magnitude for an image g(x) is defined as follows,

$$Grad = \sqrt{F_x^2 + F_y^2}$$

where '$F_x(x)$' and '$F_y(x)$' are partial derivatives in an image's horizontal and vertical direction.

Feature similarity between original ($g_1(i)$) and distorted ($g_2(i)$) images are measured in two different phases. First, the similarity map is computed locally and in the next phase, they are grouped into a single score value. It is calculated separately for phase congruency and gradient magnitude and is defined as follows,

$$FSPC\ (i) = \frac{2\text{PHASECON}_1(i).\text{PHASECON}_2(i) + C1}{\text{PHASECON}_1^2(i) + \text{PHASECON}_2^2(i) + C1}$$

where FSPC is the similarity measure between phase congruencies, 'PHASECON$_1$' and 'PHASECON$_2$' which is the phase congruencies of the original ($g_1(i)$) and distorted ($g_2(i)$) image and 'C1' is a small positive constant value used for stabilizing the resultant value.

$$FSGM(i) = \frac{2Grad_1(i).Grad_2(i) + C2}{Grad_1^2(i) + Grad_2^2(i) + C2}$$

where FSGM is the similarity measure between gradient magnitude, 'Grad$_1$' and 'Grad$_2$' which is the gradient magnitudes of original ($g_1(i)$) and distorted ($g_2(i)$) image and 'C2' is a small positive constant value used for result stabilization.

Now the similarity result of phase congruency and gradient magnitude are combined to get the local similarity values of the original ($g_1(i)$) and distorted ($g_2(i)$) image and is defined as follows,

$$LSIM(i) = [FSPC\ (i)]^\alpha + [FSGM(\ i)]^\beta$$

where 'α' and 'β' are used as parameter values to adjust the phase congruency and gradient magnitude feature values, generally, 'α' and 'β' values are set as 1.

Finally, the FSIM index is calculated as follows,

$$FSIM\ index = \frac{\sum_{i\in\Omega} LSIM(i).PCMAX(i)}{\sum_{i\in\Omega} PCMAX(i)}$$

where

$$PCMAX = \max\ (\text{PHASECON}_1(i), \text{PHASECON}_2(i))$$

Chapter 11

Case Study on Successful Digital Image Processing Application on Banana

This chapter describes a successful case study on applying digital image processing to diagnose pest and disease problems of banana. Banana is a major fruit crop that can provide abundant income to the farmers. However, this capability of banana crop is not fully enjoyed by the farmers, as there are various threats that reduce the crop's productivity. The major threat to banana production is caused by pests and diseases that reduce crop productivity leading to heavy financial loss for farmers. Symptoms of these pest and disease infections are most often found in crop leaves. In some crops, diseases are visible in the early stage; in some crops, they will be visible only in the later stage as there will be no possibility of rescuing the crop. So persistent monitoring over the plant helps identify the pest and disease in early stage and sustains the plant quality with minimized yield loss. Therefore, it becomes essential to analyze the banana leaves to identify the type of pest or disease.

If farmers can identify these infections at their earlier stage correctly, preventive measures can be taken to protect crop productivity. However, unfortunately, many farmers are unaware of the disease identification based on the plant symptom expression. It forces them to seek the support of specialists from agricultural research institutes and universities. Hence, supporting diagnostic services provided by sources such as agricultural research institutions and state farm advisory services is becoming mandatory for banana cultivators. This always involves more time consuming and needs additional cost towards advisory services. The development of an automated system requires farmers to avoid these inconveniences and have a user-friendly suggestions. Automation using advanced computer technologies like digital image processing provides feasible support to the farmers. Therefore, this chapter demonstrated the development of an accurate and automatic disease identification and classification system using image processing along with machine learning algorithm.

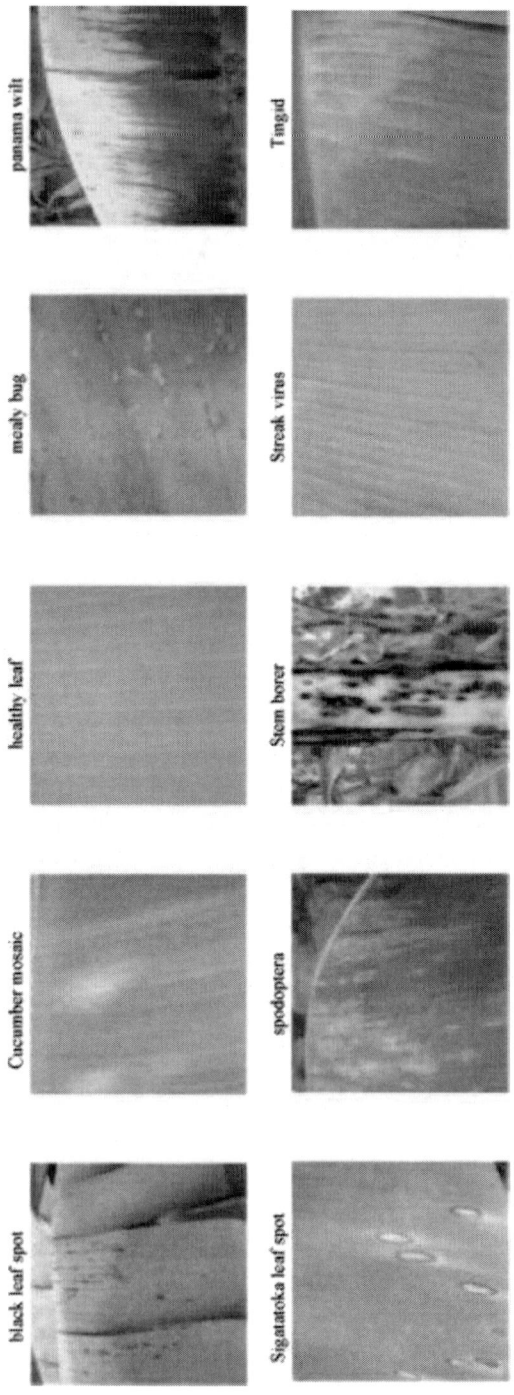

Figure 85. Sample image of diseases from banana leaf disease dataset.

11.1. Acquisition

The pest and disease symptom images are obtained from banana cv. Grand Naine from farmer fields at Thondamuthur village, Coimbatore, Tamil Nadu, India at day natural illumination condition in different times. Fifty images each for i) Sigatoka leaf spot, ii) Fusarium wilt, iii) moko wilt, iv) mealy bug, v) streak virus, vi) Tingid affected or infected leaves and vii) healthy leaves were taken as in Figure 85. The digital camera Olympus SP 510 UZ is used for image collection. The camera set up during image acquisition is at 200 ISO sensitivity, 1/125 shutter speed, 3.5 Aperture, 3072 x 2304 resolutions and in JPEG format. All the images are captured under diffusive light condition on shadow side of the canopy without direct sunlight to maintain illumination homogeneity -five images from each category are used for developing the algorithm as Calibration images; the rest of twenty-five are used as validation images.

11.2. Pre-Processing

Input leaf images are enhanced through an efficient image enhancement algorithm using the concept of fuzzy intensification operator and genetic algorithm. This proposed algorithm can enhance the contrast of leaf images taken from real time (Surya Prabha & Satheesh Kumar, 2017).

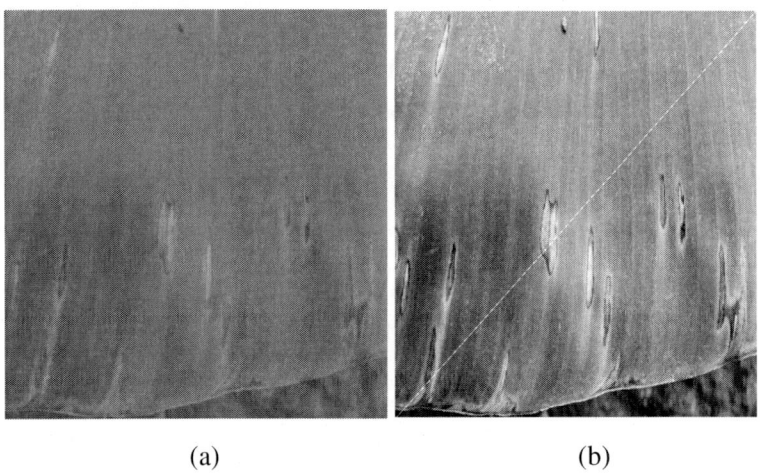

(a) (b)

Figure 86. (a) Original input image (b) after enhancement.

11.3. Segmentation

An efficient image segmentation algorithm based on the flower pollination algorithm has been applied to identify the exact infected region. This algorithm can identify the exact region of interest from an image (Prabha & Kumar, 2016).

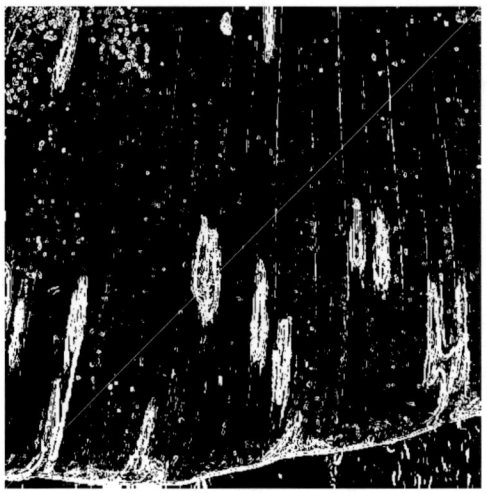

Figure 87. Segmented image.

11.4. Feature Extraction

Interest points are calculated using Features from Accelerated Segment Test (FAST) method and the Histogram of Oriented Gradients (HOG) method is used for object description to extract the features in an image.

11.5. Classification

Image classification is a higher level of image processing module which deals with the concept of machine learning algorithms and plays a vital role in object detection and recognition. So, this module plays a vital role in diagnosing the leaf image and identifying if it is healthier or infected by any diseases. It also supports accurately classifying the type of disease infection in the leaf.

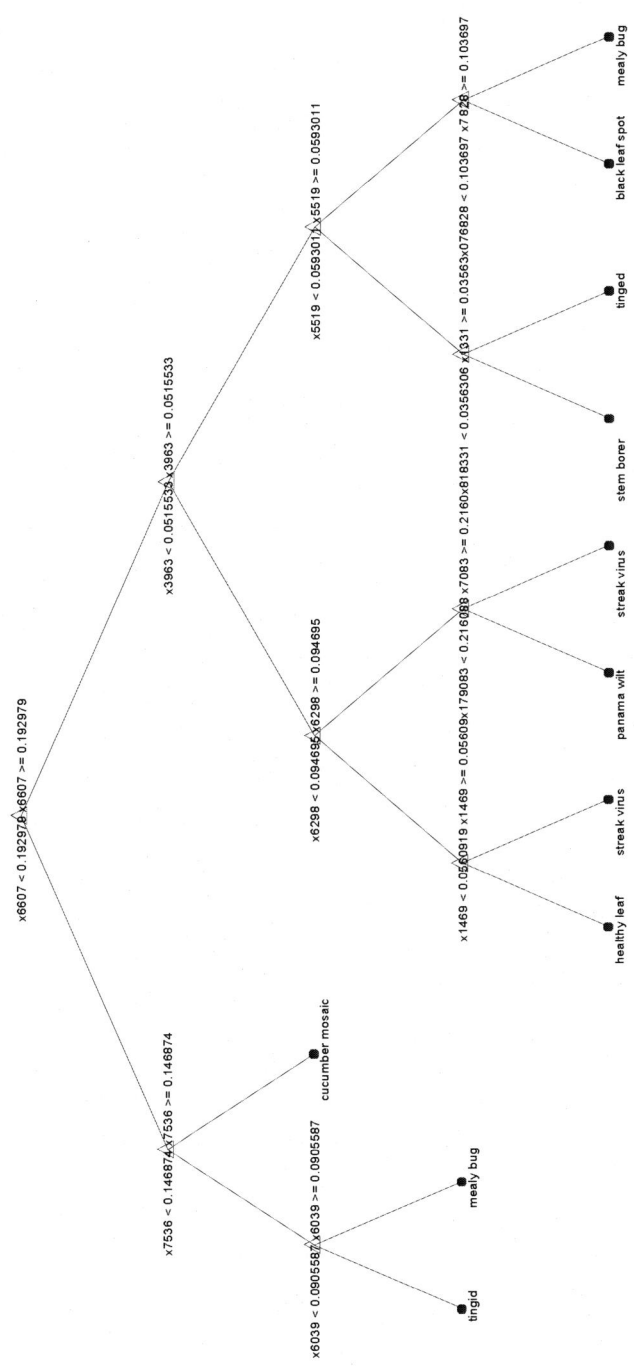

Figure 88. Output of classification tree distribution.

Figure 89. Development of GUI for banana leaf disease diagnosis.

Machine learning algorithms are computer program that learns from the training dataset and their performance is improved based on the experience it acquires through training. The decision trees algorithm has been used to classify leaf pests and diseases.

The Decision tree classifier can detect banana pests and diseases with 97.5% accuracy (Figure 89). The pest and disease diagnosis task is finally automated using the Graphical User Interface Development Environment (GUIDE). GUIDE is user-friendly. The digital image processing algorithm module is linked to the GUIDE (Figure 90). Hence, the user can instantly get the result of banana image as infected or healthy with name of the pest and disease after selecting images and pressing the ok option in GUIDE.

References

Abdul Kadir, Nugroho, L.E., Susanto, A., & Santosa, P.N. (2011). Leaf Classification Using Shape, Color, and Texture Features. *International Journal of Computer Trends and Technology*, 4, 225-230.

Arnold-Bos, A., Malkasse, J.P., & Kervern, G. (2005) A preprocessing framework for automatic underwater images denoising. *European Conference on Propagation and Systems*, Brest: France (2005), 1-8.

Avcibas, I., Sankur, B., & Sayood, K. (2002). Statistical evaluation of image quality measures. *Journal of Electronic imaging*, *11*(2), 206-223.

Baghel, M., Agrawal, S., & Silakari, S. (2012). Survey of metaheuristic algorithms for combinatorial optimization. *International Journal of Computer Applications*, *58*(19).

Bama, S.B., Valli, S.M., Raju, S., & Kumar, V.A. (2011). Content based leaf image retrieval (CBLIR) using shape, color and texture features. *Indian Journal of Computer Science and Engineering*, 2(2), 202-211.

Baterina, A.V., & Oppus, C. (2010). Image edge detection using ant colony optimization. *Wseas transactions on signal processing*, 6(2), 58-67.

Bato, P.M., Nagata, M., Cao, Q.X., Hiyoshi, K., & Kitahara, T., (2000). Study on sorting system for strawberry using machine vision (part 2): development of sorting system with direction and judgement functions for strawberry (Akihime variety). *Journal of the Japanese Society of Agricultural Machinery*. 62(2):101-110.

Bazeille, S., Quidu, I., Jaulin, L., & Malkasse, J.P. (2006). *Automatic Underwater Image Pre-Processing*. CMM'06, Brest: France, 1-8.

Beyyala, A., & Beyyala, S.P. (2012). Application for diagnosis of diseases in crops using image processing, *International Journal of LifeSciences Biotechnology & Pharma Research*, 1(2), 172-176

Bhandarkar, S.M., Zhang, Y., & Potter, W.D. (1994). An edge detection technique using genetic algorithm-based optimization. *Pattern Recognition*, 27(9), 1159-1180.

Bhattacharyya, S. (2011). A brief survey of color image preprocessing and segmentation techniques. *Journal of Pattern Recognition Research*, 1, 120-129.

Bhope, S.F., & Patil, S.P. (2010). Defects identification in textile industries. *International Journal of Chemical Sciences and Applications*, 1(1), 37-41.

Binitha, S., & Sathya, S.S. (2012). A survey of bio inspired optimization algorithms. *International journal of soft computing and engineering*, 2(2), 137-151.

Blasco, J., Aleixos, N., & Molto, E. (2007). Computer vision detection of peel defects in citrus by means of a region oriented segmentation algorithm. *Journal of Food Engineering*, 81, 535–543.

Brosnan, T., & Sun, D.W. (2004). Improving quality inspection of food products by computer vision - a review. *Journal of Food Engineering*, 61, 3–16.

Butz, P., Hofmann, C., & Tauscher, B. (2005). Recent developments in Noninvasive techniques for fresh fruit and vegetable internal quality analysis. *Journal of Food Science*, 70(9), 131-141.

Camargo, A., & Smith, J.S. (2009). An image-processing based algorithm to automatically identify plant disease visual symptoms. *Biosystems engineering*, 102(1), 9-21.

Cardoso, J.S., & Corte-Real, L. (2005). Toward a generic evaluation of image segmentation. *IEEE Transactions on Image Processing*, *14*(11), 1773-1782.

Castillo, O., Martinez-Marroquin, R., Melin, P., Valdez, F., & Soria, J. (2012). Comparative study of bio-inspired algorithms applied to the optimization of type-1 and type-2 fuzzy controllers for an autonomous mobile robot. *Information sciences*, *192*, 19-38.

Chaerle, L., & Van Der Straeten, D. (2001). Seeing is believing: imaging techniques to monitor plant health. *Biochimica et Biophysica Acta (BBA)-Gene Structure and Expression*, 1519(3), 153-166.

Charbit, M. (Ed.). (2010). *Digital signal and image processing using MATLAB* (Vol. 666). John Wiley & Sons.

Chen, J.F., Hsieh, H.N., & Do, Q.H. (2014). Predicting student academic performance: A comparison of two meta-heuristic algorithms inspired by cuckoo birds for training neural networks. *Algorithms*, 7(4), 538-553.

Civicioglu, P. (2013). Artificial cooperative search algorithm for numerical optimization problems. *Information Sciences*, *229*, 58-76.

Costa, C., Antonucci, F., Pallattino, F., Aguzzi, J., Sun, D.W., Menesatti, P. (2011). Shape analysis of agricultural products: A review of recent research advances and potential application to computer vision. *Food Bioprocess Technology*, 4, 673-692.

da Silva, J.M. (2016). Monitoring photosynthesis by in vivo chlorophyll fluorescence: application to high-throughput plant phenotyping. In *Applied Photosynthesis-New Progress*. IntechOpen.

Dadwal, M., & Banga, V.K. (2012). Color image segmentation for fruit ripeness detection: a review. *ICEECE 2012*, 190-193.

Deepa, S.N., & Sivanandam, S.N. (2011). *Principles of soft computing*. Wiley India, 745.

Dougherty, G. (2009). *Digital Image Processing for Medical Applications*. Cambridge University Press, UK. p. 440.

Duan, H., & Qiao, P. (2014). Pigeon-inspired optimization: a new swarm intelligence optimizer for air robot path planning. *International journal of intelligent computing and cybernetics*.

Ebrahimi, M.A., Mohtasebi, S.S., & Hosseinpour, R.S. (2012). Investigation of banana slices shrinkage using image processing technique. *Australian Journal of Crop Science*, 6(5):938-945

Engelbrecht, A.P. (2007). *Computational intelligence: an introduction*. John Wiley & Sons.

Fahlgren, N., Gehan, M.A., & Baxter, I. (2015). Lights, camera, action: high-throughput plant phenotyping is ready for a close-up. *Current opinion in plant biology*, 24, 93-99.

Farmer, M.E., & Jain, A.K. (2005). A wrapper-based approach to image segmentation and classification. *IEEE transactions on image processing,* 14(12), 2060-2072.

Femat-Diaz, A., Vargas-Vazquez, D., Huerta-Manzanilla, E., Rico-Garcia, E., &Herrera-Ruiz, G. (2011). Scanner image methodology (SIM) to measure dimensions of leaves for agronomical applications. *African Journal of Biotechnology*, 10(10), 1840-1847.

Fister Jr, I., Yang, X.S., Fister, I., Brest, J., & Fister, D. (2013). A brief review of nature-inspired algorithms for optimization. *arXiv preprint arXiv:1307.4186*.

References

Gandomi, A.H., & Alavi, A.H. (2012). Krill herd: a new bio-inspired optimization algorithm. *Communications in nonlinear science and numerical simulation*, *17*(12), 4831-4845.

Gao, H., Zhu, F., & Cai, J. (2010). A review of Non-destructive Detection for Fruit Quality. *IFIP Advances in Information and Communication Technology*, 317, 133-140.

Gao, H., Zhu, F., & Cai, J. (2009). A review of non-destructive detection for fruit quality. In *International Conference on Computer and Computing Technologies in Agriculture* (pp. 133-140). Springer, Berlin, Heidelberg.

Gay, P., Berruto, R., & Piccarolo, P. (2002). Fruit Color Assessment for Quality Grading Purposes. *Paper presented in ASAE Annual International Meeting/CIGR XVth World Congress,* held at Chicago, Illinois, USA during July 28-July 31, 2002. pp 1- 9.

Geman, D., Geman, S., Graffigne, C., & Dong, P. (1990). Boundary detection by constrained optimization. *IEEE Transactions on pattern analysis and machine intelligence*, *12*(7), 609-628.

Gerig, G., Kubler, O., Kikinis, R., & Jolesz, F.A. (1992). Nonlinear anisotropic filtering of MRI data. *IEEE Transactions on medical imaging*, *11*(2), 221-232.

Gilboa, G., Sochen, N., & Zeevi, Y.Y. (2002). Forward-and-backward diffusion processes for adaptive image enhancement and denoising. *IEEE transactions on image processing, 11*(7), 689-703.

Gokhale, A.M., & Yang, S. (1999). Application of Image Processing for Simulation of Mechanical Response of Multi–Length Scale Microstructures of Engineering Alloys. *Metallurgical and Materials Transactions,* 30, 2369-2381.

Gonzalez, R.C. (2009). *Digital image processing.* Pearson education India.

Greenberg, S., Aladjem, M., & Kogan, D. (2002). Fingerprint image enhancement using filtering techniques. *Real-Time Imaging*, *8*(3), 227-236.

Heath, M., Sarkar, S., Sanocki, T., & Bowyer, K. (1998). Comparison of edge detectors: a methodology and initial study. *Computer vision and image understanding*, *69*(1), 38-54.

Hore, A., & Ziou, D. (2010). Image quality metrics: PSNR vs. SSIM. In *2010 20th international conference on pattern recognition* (pp. 2366-2369). IEEE.

IMEC. (2008). *Imec launches TDI, multispectral and hyperspectral sensors.* Available at: https://optics.org/news/8/2/8.

Iqbal, K., Salam, R.A., Osman, A., & Talib, A.Z. (2007). Underwater Image Enhance-ment Using an Integrated Colour Model. *International Journal of Computer Science*, 34 (2), 1-6.

Ji, Q., & Haralick, R.M. (2002). Efficient facet edge detection and quantitative performance evaluation. *Pattern Recognition*, *35*(3), 689-700.

Jones, H.G., Stoll, M., Santos, T., Sousa, C.D., Chaves, M.M., & Grant, O.M. (2002). Use of infrared thermography for monitoring stomatal closure in the field: application to grapevine. *Journal of experimental botany,* 53(378), 2249-2260.

Jordehi, A.R. (2015). Brainstorm optimisation algorithm (BSOA): An efficient algorithm for finding optimal location and setting of FACTS devices in electric power systems. *International Journal of Electrical Power & Energy Systems*, *69*, 48-57.

Kaizua, Y., & Imou, K. (2008). A dual-spectral camera system for paddy rice seedling row detection. *Computers and Electronics in Agriculture*, 63, 49–56.

Kim, D.G., Burks, T.F., Schumann, A.W., Zekri, M., Zhao, X., & Qin, J. (2009). Detection of citrus greening using microscopic imaging. *Agricultural Engineering International*, 11, 1-17.

Ko, S.J., & Lee, Y.H. (1991). Center weighted median filters and their applications to image enhancement. *IEEE transactions on circuits and systems*, *38*(9), 984-993.

Lamb, D.W., & Brown, R.B. (2001). Remote-Sensing and Mapping of Weeds in Crops. *Journal of Agricultural Engineering Research*, 78 (2), 117-125.

Le, L., Chao, L., Xiufang, Z., & Yaozhong, P. (2008). Consistency analysis on paddy rice area survey with spot and TM images under the total restraint quantity. *The International Archives of the Photogrammetry, Remote Sensing and Spatial Information Sciences*. Vol. XXXVII. Part B8. Beijing. pp. 929-933.

Lee, S. (2011). Color Image-based Defect Detection Method and Steel Bridge Coating. *47th ASC Annual International Conference Proceedings*, 1-7.

Li, L., Zhang, Q., & Huang, D. (2014). A review of imaging techniques for plant phenotyping. *Sensors*, 14(11), 20078-20111.

Lobet, G., Pagès, L., & Draye, X. (2011). A novel image-analysis toolbox enabling quantitative analysis of root system architecture. *Plant physiology*, 157(1), 29-39.

Maia, R.D., de Castro, L.N., & Caminhas, W.M. (2012). Bee colonies as model for multimodal continuous optimization: The OptBees algorithm. In *2012 IEEE congress on evolutionary computation* (pp. 1-8). IEEE.

Maini, R., & Aggarwal, H. (2010). A Comprehensive Review of Image Enhancement Techniques. *Journal of Computing*, 2 (3), 8-13.

Mathews, J.H., & Fink, K.D. (2004). *Numerical methods using MATLAB* (Vol. 4). Upper Saddle River, NJ: Pearson prentice hall.

Mirjalili, S., Wang, G.G., & Coelho, L.D. S. (2014). Binary optimization using hybrid particle swarm optimization and gravitational search algorithm. *Neural Computing and Applications*, 25(6), 1423-1435.

Monavar, H.M., Alimardani, R., & Omid, M. (2011). Detection of red ripe tomatoes on stem using Image Processing Techniques. *Journal of American Science*, 2011;7(7), 376-379.

Moreda, G.P., Muñoz, M.A., Ruiz-Altisent, M., & Perdigones, A. (2012). Shape determination of horticultural produce using two-dimensional computer vision–A review. *Journal of Food Engineering*, 108(2), 245-261.

Murakami P.F., Turner, Michelle. R., Van den Berg, A.K., & Schaberg, P.G. (2005). *An instructional guide for leaf color analysis using digital imaging software*. Gen. Tech. Rep. NE-327. Newtown Square, PA: U.S. Department of Agriculture, Forest Service, Northeastern Research Station. p 33.

Murino, V., & Trucco, A. (2000). Three-Dimensional Image Generation and Processing in Underwater Acoustic Vision. *Proceedings of the IEEE 2000*, 88(12), 1903-1946.

Pal, S.K., & King, R. (1981). Image enhancement using smoothing with fuzzy sets. *IEEE Trans. Sys., Man, and Cyber.*, *11*(7), 494-500.

Pal, S.K., & Rosenfeld, A. (1988). Image enhancement and thresholding by optimization of fuzzy compactness. *Pattern Recognition Letters*, 7(2), 77-86.

Pass, G., Zabih, R., & Miller, J. (1997). Comparing images using color coherence vectors. In: *ACMMM*. pp. 1–14.

Patel, H.N., Jain, R.K., & Joshi, M.V. (2011). Fruit detection using improved multiple features algorithm. *International Journal of Computer Application*, 13, 1-5.

Patil, J.K., & Raj Kumar. (2011). Advances in image processing for detection of plant disease. *Journal of Advanced Bioinformatics Applications and Research*, 2(2), 135-141.

Patil, S.B., & Bodhe, S.K. (2011). Betel leaf area measurement using image processing. *International Journal on Computer Science and Engineering*, 3(7), 2656-2660.

Patil, S.B., & Bodhe, S.K. (2011). Image processing method to measure sugarcane leaf area. *International Journal of Engineering Science and Technology*, 3(8), 6394 – 6400.

Polesel, A., Ramponi, G., & Mathews, V.J. (2000). Image enhancement via adaptive unsharp masking. *IEEE transactions on image processing*, 9(3), 505-510.

Prabha, D.S., & Kumar, J.S. (2012, October). Crop disease identification using image processing methods. In *Proceedings of the National Conference on Green Computing Organized by Department of Computer Science & Research Centre*, ST Hindu College, Nagercoil, Tamil Nadu, India (pp. 5-6).

Prabha, D.S., & Kumar, J.S. (2014). Study on banana leaf disease identification using image processing methods. *International Journal of Research in Computer Science and Information Technology*, 2, 89-94.

Prabha, D.S., & Kumar, J.S. (2016). Performance evaluation of image segmentation using objective methods. *Indian J. Sci. Technol*, 9(8), 1-8.

Prabha, D.S. (2017). *Development of an efficient model for banana grading and crop disease analysis using optimized image enhancement and segmentation algorithms*. PhD Thesis submitted to Bharathiar University, Coimbatore, Tamil Nadu, India.

Prakash, S.R., & Shetty, V. (2015). Review on optimization techniques used for image compression. *International Journal of Research in Engineering and Technology*, 562-567.

Pydipati, R., Burks, T.F., & Lee., W.S. (2006). Identification of citrus disease using color texture features and discriminant analysis. *Computers and Electronics in Agriculture*, 52, 49-59.

Quevedo, R., Mendoza, F., Aguilera, J.M., Chanona, J., & Gutiérrez-López, G. (2008). Determination of senescent spotting in banana (Musa cavendish) using fractal texture Fourier image. *Journal of Food Engineering*, 84(4), 509-515.

Rajasekaran, S., & Pai, G.V. (2003). *Neural networks, fuzzy logic and genetic algorithm: synthesis and applications* (with cd). PHI Learning Pvt. Ltd.

Riyadi, S., Rahni, A.A.A., Mustafa, M.M., & Hussain, A. (2007). Shape Characteristics Analysis for Papaya Size Classification. *Paper presented in the 5th Student Conference on Research and Development –SCOReD* held at Malasia during 11-12 December 2007. pp. 1-9.

Rocha, A., Hauagge, D.C., Wainer, J., & Goldenstein, S (2010). Automatic fruit and vegetable classification from images. *Computers and Electronics in Agriculture*, 70, 96—104.

Russ, J.C., & Woods, R.P. (1995). Book Review. The Image Processing Handbook. *Journal of Computer Assisted Tomography*, 19(6), 979-981.

Sadykhov, R., Dorogush, A.V., Pushkin, Y.V., Podenok, L.P., & Ganchenko, V.V. (2007). Multispectral satellite images processing for forests and wetland regions monitoring using parallel MPI implementation. *Proc. 'Envisat Symposium 2007'*, Montreux, Switzerland 23–27 April 2007 (ESA SP-636, July 2007), 1-6.

Sakamoto, T., Gitelsonb, A.A., Nguy-Robertsonb, A.L., Arkebauerc, T.J., Wardlowb, B.D., Suykerb, A.W., Vermab, S.B., & Shibayamaa, M. (2012). An alternative method using digital cameras for continuous monitoring of crop status. *Agricultural and Forest Meteorology,* 154–155, 113–126.

Sakamoto, T., Shibayama, M., Kimura, A., & Takada, E. (2011). Assessment of digital camera-derived vegetation indices in quantitative monitoring of seasonal rice growth. *ISPRS Journal of Photogrammetry and Remote Sensing*, 66, 872–882.

Sannakki, S.S., Rajpurohit, V.S. & Arunkumar, R. (2011). A Survey on Applications of Fuzzy Logic in Agriculture. *Journal of Computer Applications,* 4 (1), 8-11.

Seng, W.C., & Mirisaee, S.H. (2009). A new method for fruits recognition system. In: *International Conference on Electrical Engineering and Informatics*, Bangi, Malaysi. pp. 130-134.

Shahzad, M., Akhter, Q., & Bibi, F. (2009). Efficient Image Enhancement Techniques. *Journal of Information & Communication Technology*, 3(1), 50-55.

Sheikh, H.R., & Bovik, A.C. (2006). Image Information and Visual Quality. *IEEE Transactions On Image Processing*, 15(2), 430-444.

Shih, F.Y. (2010). *Image processing and pattern recognition: fundamentals and techniques*. John Wiley & Sons.

Sirault, X.R., James, R.A., & Furbank, R.T. (2009). A new screening method for osmotic component of salinity tolerance in cereals using infrared thermography. *Functional Plant Biology,* 36(11), 970-977.

Solomon, C., & Breckon, T. (2011). *Fundamentals of Digital Image Processing: A practical approach with examples in Matlab*. John Wiley & Sons.

Staford, J.V. (2000). Implementing Precision Agriculture in the 21st Century. *Journal of Agricultural Engineering Research,* 76, 267-275.

Starck, J.L., Murtagh, F., Candès, E.J., & Donoho, D.L. (2003). Gray and color image contrast enhancement by the curvelet transform. *IEEE Transactions on image processing*, 12(6), 706-717.

Surya Prabha, D., & Satheesh Kumar, J. (2017). An efficient image contrast enhancement algorithm using genetic algorithm and fuzzy intensification operator. *Wireless Personal Communications,* 93(1), 223-244.

Suryaprabha, D & Satheeshkumar, J. (2021). Crop disease identification using image processing methods. In *Proceedings of the National Conference on Green Computing organized by Department of Computer Science & Research Centre*, ST Hindu College, Nagercoil, Tamil Nadu, India, pages 174–179, 2012.

Suryaprabha, D., Satheeshkumar, J., & Seenivasan, N. (2020). Classical and Fuzzy Based Image Enhancement Techniques for Banana Root Disease Diagnosis: A Review and Validation. *Oriental Journal of Computer Science and Technology*, 13(1), 50-62.

Thomas, P. (2009). *Remote sensing and satellite image processing*. SECON Private Limited, Bangalore, India. pp. 9.

References

Tucker, C.J. (1979). Red and photographic infrared linear combinations for monitoring vegetation. *Remote sensing of Environment,* 8(2), 127-150.

Vajda, F. (1994). Techniques and trends in digital image processing and computer vision. *IEEE Colloquium on Mathematical Modelling and Simulation of Industrial and Economic Processes,* 1.

Vince, J., & Vince, J.A. (2010). *Mathematics for computer graphics* (Vol. 3). London: Springer.

Wang, Z., Bovik, A.C., Sheikh, H.R., & Simoncelli, E.P. (2004). Image quality assessment: from error visibility to structural similarity. *IEEE transactions on image processing,* 13(4), 600-612.

Yang, C.C., Prasher, S.O., Landry, J.A., Perret, J., & Ramaswamy, H.S. (2000). Recognition of weeds with image processing and their use with fuzzy logic for precision farming. *Canadian Agricultural Engineering,* 42 (4), 195-200.

Yang, X.S. (2012). Flower pollination algorithm for global optimization. In *International conference on unconventional computing and natural computation* (pp. 240-249). Springer, Berlin, Heidelberg.

Yitzhaky, Y., & Peli, E. (2003). A method for objective edge detection evaluation and detector parameter selection. *IEEE Transactions on pattern analysis and machine intelligence,* 25(8), 1027-1033

Zelelew, H.M., Papagiannakis, A.T., & Masad, E. (2008). Application of digital image processing techniques for asphalt concrete mixture images. *The 12th International Conference of International Association for Computer Methods and Advances in Geomechanics (IACMAG),* Goa, India, pp.119-124.

Zhang, H., Fritts, J.E., & Goldman, S.A. (2008). Image segmentation evaluation: A survey of unsupervised methods. *Computer Vision and Image Understanding,* 110 (2), 260-280.

Zhang, L., Zhang, L., Mou, X., & Zhang, D. (2011). FSIM: a feature similarity index for image quality assessment. *IEEE transactions on Image Processing,* 20(8), 2378-2386.

Zhang, Y.J. (1996). A survey on evaluation methods for image segmentation. *Pattern recognition,* 29(8), 1335-1346.

Zhang, Y.J. (1997). Evaluation and comparison of different segmentation algorithms. *Pattern recognition letters,* 18(10), 963-974.

Zheng, X., & Wang, X. (2010). Leaf vein extraction based on gray-scale morphology. *International Journal Image, Graphics and Signal Processing,* 2, 25-31

Zhu, S.Y., Plataniotis, K.N., & Venetsanopoulos, A.N. (1999). Comprehensive analysis of edge detection in color image processing. *Optical Engineering,* 38(4), 612-625.

Index

A

accuracy, 11, 12, 14, 15, 16, 17, 21, 23, 33, 34, 37, 38, 41, 46, 57, 62, 94, 95, 104, 136, 137, 139, 140, 141, 145, 157
acquisition methods, 5, 51
active contour, 77, 91, 92, 123
agricultural, viii, 22, 23, 30, 51, 54, 123, 129, 131, 151, 159, 160, 162, 164, 165, 171, 172
agriculture sector, vii, viii, 11, 21, 22, 23
algorithm(s), viii, 4, 9, 13, 17, 23, 24, 25, 37, 38, 41, 44, 46, 47, 52, 53, 77, 83, 88, 89, 98, 99, 104, 105, 106, 108, 109, 110, 111, 112, 114, 116, 117, 118, 127, 129, 131, 132, 133, 136, 137, 139, 140, 142, 145, 151, 153, 154, 157, 159, 160, 161, 162, 163, 164, 165, 171
analysis, vii, viii, 1, 2, 3, 7, 8, 12, 14, 16, 17, 19, 20, 23, 25, 26, 29, 30, 33, 34, 36, 39, 41, 43, 49, 51, 52, 53, 67, 69, 133, 146, 147, 159, 160, 161, 162, 163, 165, 171
analytical, 135, 136, 137
ant colony optimization (ACO), 110, 112, 113, 159
area, 9, 13, 14, 29, 30, 34, 36, 39, 46, 54, 117, 123, 126, 127, 139, 141, 162, 163, 172
assessment, 5, 15, 30, 31, 38, 135, 136, 143, 145, 147, 161, 164, 165, 171

B

banana, viii, 48, 123, 127, 130, 151, 152, 153, 156, 157, 160, 163, 164, 171, 173
based, vii, 4, 6, 7, 9, 17, 22, 23, 26, 28, 33, 34, 38, 41, 43, 44, 46, 48, 54, 57, 58, 64, 65, 69, 72, 77, 78, 79, 82, 83, 84, 85, 87, 88, 89, 90, 91, 93, 94, 95, 98, 99, 100, 101, 102, 103, 104, 105, 106, 107, 110, 112, 114, 117, 118, 119, 120, 121, 129, 131, 132, 135, 136, 137, 138, 139, 140, 143, 145, 146, 147, 151, 154, 157, 159, 160, 162, 164, 165
boundary, 8, 33, 34, 36, 37, 38, 39, 44, 46, 47, 77, 91, 92, 126, 161

C

camera, 3, 8, 40, 51, 52, 53, 54, 153, 160, 161, 164
characteristic(s), 16, 30, 32, 33, 37, 87, 104, 112, 117, 129, 136, 137, 138, 139, 143, 145, 163
classification, 5, 8, 13, 14, 17, 24, 31, 32, 33, 34, 35, 36, 37, 38, 85, 88, 104, 123, 128, 129, 131, 132, 151, 154, 155, 159, 160, 163
classifiers, 37, 48, 132
clustering, 8, 9, 14, 24, 43, 44, 87, 88, 89, 141, 142
component, 25, 26, 34, 35, 49, 53, 74, 89, 94, 101, 106, 115, 116, 133, 164
crop(s), vii, 22, 23, 25, 26, 28, 29, 30, 39, 52, 53, 54, 151, 159, 160, 162, 163, 164, 171

curve, 37, 46, 91, 92, 120, 138, 139, 140, 141

D

derivative, 72, 79, 81, 82, 96, 97, 99, 100, 103, 104
description, 36, 46, 62, 154
descriptors, 33, 36, 37, 46, 48, 123, 126, 128
detection, 7, 8, 13, 16, 17, 19, 20, 24, 43, 52, 53, 59, 72, 78, 79, 81, 83, 94, 95, 96, 99, 100, 101, 104, 105, 109, 110, 117, 131, 154, 159, 160, 161, 162, 163, 165
digital image processing, vii, viii, 1, 11, 41, 151, 157, 160, 161, 164, 165

E

edge, 5, 7, 8, 16, 17, 19, 43, 45, 58, 59, 66, 69, 72, 74, 77, 78, 79, 81, 83, 91, 93, 94, 95, 96, 99, 100, 101, 102, 103, 104, 105, 108, 109, 110, 111, 113, 117, 120, 121, 139, 140, 141, 159, 161, 165
empirical, 135, 136, 137
enhanced, 5, 6, 16, 17, 19, 20, 41, 43, 57, 58, 59, 61, 62, 63, 65, 66, 70, 72, 73, 74, 99, 104, 107, 108, 109, 145, 146, 153
enhancement, vii, viii, 2, 4, 13, 14, 16, 17, 18, 20, 22, 30, 41, 51, 57, 61, 62, 63, 64, 65, 66, 67, 69, 72, 73, 74, 78, 79, 106, 136, 137, 153, 161, 162, 163, 164, 171
entropy, 36, 37, 120, 123, 142, 143, 144
evaluation, 15, 30, 34, 98, 106, 112, 135, 136, 137, 138, 139, 140, 141, 143, 144, 145, 159, 160, 161, 163, 165
extraction, vii, viii, 8, 13, 14, 17, 34, 37, 46, 123, 126, 128, 129, 130, 154, 165

F

farming, 26, 27, 28, 30, 39, 165, 173
feature(s), vii, viii, 2, 4, 8, 9, 12, 13, 14, 16, 17, 24, 25, 26, 30, 32, 33, 34, 37, 38, 41, 44, 46, 49, 51, 52, 67, 88, 94, 123, 126, 128, 129, 130, 132, 133, 137, 145, 146, 147, 148, 149, 154, 159, 163, 165
filter(s), 5, 18, 58, 66, 67, 68, 69, 70, 72, 73, 83, 162
filtering, 4, 5, 6, 19, 24, 51, 58, 66, 68, 69, 70, 71, 73, 74, 75, 78, 79, 161
fluorescence, 51, 54, 160
formation, 2, 3, 12, 41
fruit(s), viii, 21, 23, 24, 25, 26, 34, 35, 36, 37, 38, 43, 46, 48, 53, 126, 151, 159, 160, 161, 163, 164, 171
fundamental, viii, 5, 41, 57
fuzzy-based, 65

G

genetic algorithms (GA), 106, 109, 110, 111
gradient(s), 43, 69, 72, 79, 81, 83, 91, 93, 94, 95, 96, 100, 101, 102, 103, 104, 106, 108, 119, 120, 123, 128, 129, 145, 147, 148, 149, 154
gray level, 5, 33, 57, 59, 60, 62, 66, 130
growth, 22, 39, 84, 94, 164

H

hierarchical, 89
histogram, 8, 18, 20, 34, 35, 36, 43, 44, 59, 62, 63, 64, 65, 86, 87, 93, 94, 95, 123, 128, 129, 130, 154
hybrid, 93, 95, 105, 162
hyperspectral, 51, 53, 115, 161

I

identification, vii, 15, 20, 22, 23, 24, 32, 34, 44, 46, 108, 147, 151, 159, 163, 164, 173
index, 5, 88, 141, 145, 146, 147, 149, 165
industries, vii, 16, 19, 159

L

land, 13, 22, 26, 28, 30, 31

Index

M

management, 2, 9, 13, 22, 26, 27, 28, 29, 30, 54, 173
measure(s), 7, 9, 28, 31, 36, 37, 38, 39, 43, 46, 48, 49, 53, 54, 85, 88, 90, 120, 123, 124, 125, 136, 137, 138, 139, 140, 141, 142, 143, 144, 145, 146, 147, 149, 151, 159, 160, 163
measurement, 7, 8, 9, 23, 33, 39, 54, 79, 81, 90, 126, 127, 139, 141, 146, 163
metaheuristic, 109, 110, 117, 159
method(s), vii, 7, 8, 10, 11, 14, 15, 16, 17, 18, 19, 20, 22, 23, 24, 25, 32, 33, 34, 36, 38, 39, 41, 43, 44, 45, 46, 47, 48, 49, 57, 58, 59, 60, 61, 62, 63, 64, 65, 66, 67, 68, 69, 70, 71, 72, 73, 74, 77, 78, 79, 80, 81, 82, 83, 84, 85, 86, 87, 88, 89, 90, 91, 92, 93, 94, 95, 96, 97, 98, 99, 100, 101, 102, 103, 104, 106, 108, 109, 110, 111, 112, 128, 129, 132, 135, 136, 137, 138, 139, 140, 141, 142, 143, 144, 145, 146, 147, 154, 162, 163, 164, 165, 171
metrics, 5, 135, 161
modules, 2, 11, 12, 20, 24, 41, 42, 62, 65
morphological, 24, 32, 36, 44, 45, 46, 52
multispectral, 51, 53, 161, 164

N

neighborhood, 5, 57, 58, 66, 67, 69, 70, 79, 88, 103, 104, 115, 116, 117, 145
Newton-Raphson, 100, 101, 102, 104
non-reference, 5, 135

O

objective, 92, 96, 97, 98, 99, 100, 101, 102, 103, 109, 117, 121, 135, 136, 163, 165
optimization, 95, 96, 97, 98, 99, 100, 103, 104, 106, 109, 110, 112, 113, 114, 117, 132, 159, 160, 161, 162, 163, 165

P

paddy crop, 39
particle swarm optimization (PSO), 110, 114, 115, 116, 162
partitional, 88
pattern, viii, 33, 37, 38, 48, 49, 100, 128, 131, 139, 159, 161, 162, 164, 165
pattern recognition, viii, 33, 49, 131, 139, 159, 161, 162, 164, 165
plant disease, 22, 23, 24, 25, 160, 163
precision, 21, 22, 26, 27, 28, 30, 39, 139, 140, 164, 165
pre-processing, 67, 153, 159
Prewitt edge method, 7, 65, 79, 81, 82, 148
problem(s), vii, viii, 21, 32, 38, 54, 61, 65, 77, 91, 93, 95, 97, 98, 99, 100, 104, 109, 110, 111, 112, 114, 115, 116, 117, 119, 120, 131, 132, 136, 151, 160

Q

quality, vii, viii, 7, 14, 15, 16, 18, 19, 20, 22, 25, 26, 34, 38, 39, 41, 43, 51, 54, 57, 59, 68, 72, 126, 135, 138, 140, 141, 145, 146, 147, 151, 159, 161, 164, 165, 171

R

receiver, 28, 138, 139
reference, 5, 49, 64, 126, 127, 135, 136, 137
region(s), 2, 8, 9, 15, 23, 24, 33, 35, 36, 39, 43, 44, 47, 51, 54, 60, 61, 63, 64, 66, 69, 72, 77, 78, 79, 81, 84, 85, 87, 88, 89, 90, 91, 92, 93, 94, 95, 103, 104, 108, 109, 111, 113, 120, 123, 126, 129, 143, 144, 145, 154, 159, 164
regional, 34, 36, 37, 46, 77, 126
remote sensing, 13, 14, 30, 31, 39, 51, 53, 54, 89, 162, 164, 165
Roberts edge method, 7, 59, 79, 80, 81

S

sector(s), vii, 11, 12, 13, 15, 16, 18, 20, 21, 23, 26, 30, 38, 51, 53, 54, 89
segmentation, vii, viii, 8, 14, 16, 17, 18, 22, 23, 24, 34, 36, 43, 52, 53, 62, 65, 77, 78,

84, 85, 87, 89, 90, 91, 93, 94, 95, 101, 136, 137, 138, 139, 140, 141, 142, 143, 144, 145, 154, 159, 160, 163, 165, 171
shape measure, 145
sharpening, 4, 5, 6, 14, 58, 59, 66, 67, 72, 73
signal, 10, 13, 18, 53, 74, 145, 159, 160, 165
similarity, 77, 87, 88, 90, 136, 137, 141, 142, 143, 145, 146, 147, 148, 149, 165
smoothing, 4, 6, 18, 44, 58, 67, 68, 70, 71, 72, 73, 81, 84, 106, 162
Sobel edge method, 7, 59, 79, 81, 148
structural, 20, 111, 117, 145, 146, 147, 165
subjective, 135, 136
successful, 5, 52, 64, 112, 117, 151

T

techniques, vii, 4, 9, 13, 14, 16, 17, 18, 22, 23, 24, 30, 34, 38, 41, 44, 54, 57, 66, 70, 73, 93, 98, 99, 100, 104, 106, 109, 110, 111, 112, 123, 129, 131, 135, 138, 145, 159, 160, 161, 162, 163, 164, 165, 171
thermography, 51, 54, 161, 164
thresholding, 8, 16, 17, 36, 43, 44, 77, 78, 79, 83, 85, 86, 87, 104, 162
tomography, 51, 54, 163
transformation, 5, 6, 7, 17, 25, 41, 47, 57, 58, 59, 60, 61, 62, 63, 74

V

vector, 17, 22, 24, 33, 38, 49, 90, 91, 93, 94, 99, 100, 101, 102, 103, 106, 120, 129, 131, 132, 145, 147
vision system, 19, 49, 51, 52
visualization, 2, 7

About the Authors

Suryaprabha Deenan

Dr. Suryaprabha Deenan is an Assistant Professor at the Department of Information Technology, Nehru Arts and Science College, Coimbatore, Tamil Nadu, India. She graduated with B.Com with a Computer Application specialization from Madurai Kamaraj University, where she also obtained M.C.A. She did her M.phil and PhD at Bharathiar University, Coimbatore, India. She worked as Post-Doctoral Fellow and Teaching Assistant at Tamil Nadu Agricultural University; and as Assistant Professor at Karpagam University, Coimbatore. She has taught 12 undergraduate courses. Her research interests focused on image processing techniques and Artificial Intelligence in Agriculture. She developed an AI model for diagnosis and damage assessment of nematode-infected banana roots, an efficient segmentation method, tool development for banana fruit quality analysis, an efficient model for banana grading, crop disease analysis using optimized image enhancement and segmentation algorithms, and automated fall armyworm early warning expert system Using artificial intelligence-driven internet of things. She is the author of 13 referenced research papers, one book and two chapters.

About the Authors

Satheeshkumar Janakiraman

Dr. J. Satheesh Kumar is an Associate Professor in the Department of Computer Applications, School of Computer Science and Engineering, Bharathiar University, Coimbatore, Tamil Nadu, India. He has 19 plus years of research and teaching experience. His area of specialization includes soft computing, networks, Image processing, and medical imaging. He obtained five research grants from UGC- New Delhi, IGCAR-Kalpakkam, TN-State Plan, and DST-ICPS. He is the author of 72 referenced international research papers, 17 national papers, and 8 book chapters.

Seenivasan Nagachandrabose

Dr. Seenivasan Nagachandrabose is Professor at the Department of Nematology, Tamil Nadu Agricultural University, Coimbatore, Tamil Nadu, India. He graduated with a BSc Agriculture from Tamil Nadu Agricultural University, where he also obtained MSc and PhD in Plant Nematology. After a few years of postdoctoral research, he was appointed to Tamil Nadu Agricultural University in 2004. He has taught eight undergraduate and three post-graduate courses. His research interests focused on biological control of nematodes, integrated management strategies, nematode resistance in *Musa* spp., entomopathogenic nematodes and molecular identification of nematodes. He developed nematode management strategies in rice, banana, cotton and medicinal coleus to benefit the farming community. He developed nematode management strategies in rice, banana, cotton and medicinal coleus to benefit the farming community. He was trained on metabolic markers at Michigan State University, USA in 2009-10. He got young scientist award-2012 from Science and Engineering Research Council, DST, New Delhi. He was awarded Ramen Fellowship by University Grant Commission (UGC), New Delhi, India to pursue postdoctoral research at North Dakota State University, Fargo, USA in 2016-17. He also received Research Award-2016 from UGC, India. He obtained four research grants from the National Medicinal Plants Board, India; Department of Science and Technology, India; Life Science Research Board, DRDO, India; and UGC, India. He is the author of 66 referenced research papers, five books and three-book chapters.